GENERAL CLASS
AMATEUR RADIO
FCC TEST MANUAL

by

James Lyons
Richard Mintz
Martin Schwartz

edited by
Donna Bates
Marie Piepol

Published by
AMECO CORPORATION
224 EAST SECOND STREET
MINEOLA, NEW YORK 11501

GENERAL CLASS AMATEUR RADIO FCC TEST MANUAL

Copyright 1994
by the
Ameco Corporation

All rights reserved. This book, or parts thereof, may not be reproduced in any form, without the written permission of the publisher.

Library of Congress Catalog No. 90-81316

ISBN No. 0-912146-28-1

Printed in the United States of America

TABLE OF CONTENTS

Preface 4

General Class Syllabus 5

Commission's Rules 7

Operating Procedures 21

Radio Wave Propagation 32

Amateur Radio Practices 43

Electrical Principles 61

Circuit Components 72

Practical Circuits 76

Signals and Emissions 81

Antennas and Feedlines 87

Sample General Class Examination 103

RST REPORTING SYSTEM. 110

FREQUENCY CHART 111

TABLE OF EMISSIONS. 112

FCC EMISSION DEFINITIONS 114

ANTENNA ELEMENT EQUATIONS 116

GLOSSARY. 117

PREFACE

This General Class FCC Test Manual is part of a series of books published by the Ameco Corporation for the purpose of preparing individuals for the Federal Communications Commission's Amateur Radio Operator's examinations.

The questions and multiple choice answers in this manual have been issued by the Volunteer Examiner Coordinator's Committee under the supervision of the FCC. The questions and multiple choice answers on the actual examination will be drawn, word for word, from these questions.

The questions in this book are divided into subelements. A specific number of questions (shown under each subelement heading) will be taken from each subelement. In those instances where the authors feel that the correct multiple choice answer is complete and adequate for the proper understanding of the subject matter, there is little or no discussion; only the correct answer is indicated. In most of the questions, the discussions explain the correct multiple choice answers and give additional useful material that helps with the understanding of the questions and answers.

Periodically the Federal Communications Commission changes some of the Rules and Regulations of Part 97. All of the changes that affect the General Class examinations have been incorporated into this manual. However, the General Class examinations that are given prior to July 1, 1994 will not include questions that have been affected by the new rules. After July 1, 1994 the General Class tests may include the new questions. Regardless of when the test is taken, the prospective amateur operator can study this manual without being concerned about the recent changes.

The FCC/VEC examination for the General Class license consists of 25 questions. The minimum passing score is 19 questions answered correctly.

Although this guide deals with Element 3 of the Amateur Radio Operator's examinations, the Ameco Corporation publishes guides covering all the other amateur elements as well as code courses for learning International Morse Code. If additional theory background information is required, it is suggested that the Amateur Radio Theory Course (Cat.#102-01) be consulted. See back cover for details.

It is the authors' firm belief that an amateur radio operator will enjoy the hobby much more and will be a better operator if he or she has a good understanding of basic electronics. GOOD LUCK!

James Lyons
Richard Mintz
Martin Schwartz

GENERAL CLASS SYLLABUS
QUESTION POOL
ELEMENT 3B (GENERAL CLASS)

as released by Question Pool Committee
National Conference of
Volunteer Examiner Coordinators
December 1, 1993

SUBELEMENT G1 - COMMISSION'S RULES [4 exam questions - 4 groups]

G1A General control operator frequency privileges; local control, repeater and harmful interference definitions, third-party communications
G1B Antenna structure limitations; good engineering and good amateur practice; beacon operation; restricted operation; retransmitting radio signals
G1C Transmitter power standards; type acceptance of external RF-power amplifiers; standards for type acceptance of external RF-power amplifiers; HF data emission standards
G1D Examination element preparation; examination administration; temporary station identification

SUBELEMENT G2 - OPERATING PROCEDURES [3 exam questions - 3groups]

G2A Phone, RTTY, repeater, VOX and full break-in CW
G2B Operating courtesy, antenna orientation and HF operations, including logging practices, ITU Regions
G2C Emergencies, including drills, communications and amateur auxiliary to FOB

SUBELEMENT G3 - RADIO WAVE PROPAGATION [3 exam questions - 3 groups]

G3A Ionospheric disturbances; sunspots and solar radiation
G3B Maximum usable frequency, propagation "hops"
G3C Height of ionospheric regions, critical angle and frequency, HF scatter

SUBELEMENT G4 - AMATEUR RADIO PRACTICES [5 exam questions - 5 groups]

G4A Two-tone test: electronic TR switch, amplifier neutralization
G4B Test equipment; oscilloscope; signal tracer; antenna noise bridge; monitoring oscilloscope; field-strength meters
G4C Audio rectification in consumer electronics, RF ground
G4D Speech processors; PEP calculations; wire sizes and fuses
G4E RF safety

SUBELEMENT G5 - ELECTRICAL PRINCIPLES [2 exam questions - 2 groups]

G5A Impedance, including matching, resistance, including ohm, reactance, inductance, capacitance and metric divisions of these values
G5B Decibel, Ohm's Law, current and voltage dividers, electrical power calculations and series and parallel components, transformers (either voltage or impedance), sine wave root-mean-square (RMS) value

SUBELEMENT G6 - CIRCUIT COMPONENTS [1 exam question - 1 group]

G6A Resistors, capacitors, inductors, rectifiers and transistors, etc.

SUBELEMENT G7 - PRACTICAL CIRCUITS [1 exam question - 1 group]

G7A Power supplies and filters, single-sideband transmitters and receivers

SUBELEMENT G8 - SIGNALS AND EMISSIONS [2 exam questions - 2 groups]

G8A Signal information, AM, FM, single and double sideband and carrier, bandwidth, modulation envelope, deviation, overmodulation
G8B Frequency mixing, multiplication, bandwidths, HF data communications

SUBELEMENT G9 - ANTENNAS AND FEED LINES [4 exam questions - 4 groups]

G9A Yagi antennas, physical dimensions, impedance matching, radiation patterns, directivity and major lobes
G9B Loop antennas, physical dimensions, impedance matching, radiation patterns, directivity and major lobes
G9C Random wire antennas, physical dimensions, impedance matching, radiation patterns, directivity and major lobes; feedpoint impedance of 1/2-wavelength dipole and 1/4-wavelength vertical antennas
G9D Popular antenna feed lines, characteristic impedance and impedance matching; SWR calculations

SUBELEMENT G1
COMMISSION'S RULES
[4 exam questions - 4 groups]

G1A General control operator frequency privileges; local control, repeater and harmful interference definitions, third-party communications

G1A01 What are the frequency limits for General class operators in the 160-meter band? [97.301d]*
A. 1800 - 1900 kHz
B. 1900 - 2000 kHz
C. 1800 - 2000 kHz
D. 1825 - 2000 kHz
 The answer is C. A General class operator may operate CW, image, phone, RTTY or data in the entire band.

G1A02 What are the frequency limits for General class operators in the 75/80-meter band (ITU Region 2)? [97.301d]
A. 3525 - 3750 kHz and 3850 - 4000 kHz
B. 3525 - 3775 kHz and 3875 - 4000 kHz
C. 3525 - 3750 kHz and 3875 - 4000 kHz
D. 3525 - 3775 kHz and 3850 - 4000 kHz
 The answer is A. General class license holders may operate CW, phone or image from 3.85 to 4.0 MHz. He may operate CW, RTTY or data from 3.525 to 3.750 MHz.

G1A03 What are the frequency limits for General class operators in the 40-meter band (ITU Region 2)? [97.301d]
A. 7025 - 7175 kHz and 7200 - 7300 kHz
B. 7025 - 7175 kHz and 7225 - 7300 kHz
C. 7025 - 7150 kHz and 7200 - 7300 kHz
D. 7025 - 7150 kHz and 7225 - 7300 kHz
 The answer is D. General class license holders may operate CW, RTTY or data from 7025 to 7150 kHz. He may operate CW, phone or image from 7225 to 7300 kHz.

G1A04 What are the frequency limits for General class operators in the 30-meter band? [97.301d]

---

* The number in the brackets at the end of the questions in this subelement (Commission's Rules) indicates the section and paragraph in the Rules and Regulations of Part 97 Amateur Radio Service where information concerning the question can be found.

A. 10100 - 10150 kHz
B. 10100 - 10175 kHz
C. 10125 - 10150 kHz
D. 10125 - 10175 kHz

The answer is A. The General class operator is permitted to transmit only CW, RTTY and data from 10100 to 10150 kHz.

G1A05 What are the frequency limits for General class operators in the 20-meter band? [97.301d]
A. 14025 - 14100 kHz and 14175 - 14350 kHz
B. 14025 - 14150 kHz and 14225 - 14350 kHz
C. 14025 - 14125 kHz and 14200 - 14350 kHz
D. 14025 - 14175 kHz and 14250 - 14350 kHz

The answer is B. The General class operator may operate CW, RTTY or data from 14.025 to 14.150 MHz. He may operate CW, phone or image from 14.225 to 14.350 MHz.

G1A06 What are the frequency limits for General class operators in the 15-meter band? [97.301d]
A. 21025 - 21200 kHz and 21275 - 21450 kHz
B. 21025 - 21150 kHz and 21300 - 21450 kHz
C. 21025 - 21150 kHz and 21275 - 21450 kHz
D. 21025 - 21200 kHz and 21300 - 21450 kHz

The answer is D. General class operators can operate CW, RTTY or data from 21.025 to 21.2 MHz. He may operate CW, phone or image from 21.3 to 21.45 MHz.

G1A07 What are the frequency limits for General class operators in the 12-meter band? [97.301d]
A. 24890 - 24990 kHz
B. 24890 - 24975 kHz
C. 24900 - 24990 kHz
D. 24900 - 24975 kHz

The answer is A. The General class license holder may operate CW, RTTY or data from 24890 to 24930 kHz. He may operate CW, phone or image from 24930 to 24990 kHz.

G1A08 What are the frequency limits for General class operators in the 10-meter band? [97.301d]
A. 28000 - 29700 kHz
B. 28025 - 29700 kHz
C. 28100 - 29600 kHz
D. 28125 - 29600 kHz

The answer is A. The General class operator may operate CW, RTTY or data from 28000 to 28300 kHz. He may operate CW, phone or image from 28.3 to 29.7 MHz.

G1A09 What are the frequency limits within the 160-meter band for phone emissions? [97.305c]
A. 1800 - 2000 kHz
B. 1800 - 1900 kHz
C. 1825 - 2000 kHz
D. 1825 - 1900 kHz

The answer is A. Phone emissions may be used over the entire 160-meter wavelength band (1800 - 2000 kHz). See appendixes 3 and 4 for the definition of "phone" and the definitions of the other terms that are used to indicate the various emission types.

G1A10 What are the frequency limits within the 80-meter band in ITU Region 2 for CW emissions? [97.305a]
A. 3500 - 3750 kHz
B. 3700 - 3750 kHz
C. 3500 - 4000 kHz
D. 3890 - 4000 kHz

The answer is C. The meaning of "CW" is "continuous wave". It indicates International Morse Code telegraphy emissions. It may be used over the entire 80-meter wavelength band (3500 - 4000 kHz). See appendixes 3 and 4.

G1A11 What are the frequency limits within the 40-meter band in ITU Region 2 for image emissions? [97.305c]
A. 7225 - 7300 kHz
B. 7000 - 7150 kHz
C. 7100 - 7150 kHz
D. 7150 - 7300 kHz

The answer is D. The term "image" indicates facsimile and television emissions. They may be used in the 7150 to 7300 kHz part of the 40-meter wavelength band in ITU Region 2. See appendixes 3 and 4.

G1A12 What are the frequency limits within the 30-meter band for RTTY emissions? [97.305c]
A. 10125 - 10150 kHz
B. 10125 - 10140 kHz
C. 10100 - 10150 kHz
D. 10100 - 10140 kHz

The answer is C. The term "RTTY" stands for narrow-band, direct-printing telegraphy emissions. RTTY may be used over the entire 30-meter wavelength band (10100 - 10150 kHz) that is available for amateur use.

G1A13 What are the frequency limits within the 20-meter band for

image emissions? [97.305c]
A. 14025 - 14300 kHz
B. 14150 - 14350 kHz
C. 14025 - 14350 kHz
D. 14150 - 14300 kHz
  The answer is B. See answer G1A11 and appendixes 3 and 4.

G1A14 What are the frequency limits within the 15-meter band for image emissions? [97.305c]
A. 21250 - 21300 kHz
B. 21150 - 21450 kHz
C. 21200 - 21450 kHz
D. 21100 - 21300 kHz
  The answer is C. See answer G1A11 and appendixes 3 and 4.

G1A15 What are the frequency limits within the 12-meter band for phone emissions? [97.305c]
A. 24890 - 24990 kHz
B. 24890 - 24930 kHz
C. 24930 - 24990 kHz
D. Phone emissions are not permitted in this band
  The answer is C. The term "PHONE" means speech or other sound emissions having designators as listed in Appendix 3 and 4 and can be transmitted from 24930 to 24990 kHz.

G1A16 What are the frequency limits within the 10-meter band for phone emissions? [97.305c]
A. 28000 - 28300 kHz
B. 29000 - 29700 kHz
C. 28300 - 29700 kHz
D. 28000 - 29000 kHz
  The answer is C. Phone emissions can be transmitted on the 10-meter band from 28300 to 29700 kHz

G1A17 As a General class control operator at a Novice station, how must you identify your station when transmitting on 7250 kHz? [97.119d]
A. With your call sign, followed by the word "controlling" and the Novice call sign
B. With the Novice call sign, followed by the slant bar "/" (or any suitable word) and your own call sign
C. With your call sign, followed by the slant bar "/" (or any suitable word) and the Novice call sign
D. A Novice station should not be operated on 7250 kHz, even with a General control operator

The answer is B. The General class operator may be the control operator on this frequency and since his privileges exceed that of the station licensee, he must give his own call sign after that of the Novice station's call sign. This is in accordance with paragraph 97.119d of the FCC Rules and Regulations.

G1A18 Under what circumstances may a 10-meter repeater retransmit the 2-meter signal from a Technician class operator? [97.205a]
A. Under no circumstances
B. Only if the station on 10 meters is operating under a Special Temporary Authorization allowing such retransmission
C. Only during an FCC-declared general state of communications emergency
D. Only if the 10-meter control operator holds at least a General class license

The answer is D. Although the Technician class operator may not operate on the 10-meter band his 2-meter signal may be retransmitted on 10-meters, provided that the control operator of the 10-meter transmitter has frequency privileges in this band. The important point to remember is that the control operator of the transmitting station must have the proper license.

G1A19 What kind of amateur station automatically retransmits the signals of other stations? [97.3a35]
A. Repeater station
B. Space station
C. Telecommand station
D. Relay station

The answer is A. A repeater is a receiver-transmitter device that receives a signal on one frequency and automatically retransmits it on another frequency. They are located on hills or other high points and can be used to extend the range of low-power hand held and mobile stations.

G1A20 What name is given to a form of interference that seriously degrades, obstructs or repeatedly interrupts a radiocommunication service? [97.3a21]
A. Intentional interference
B. Harmful interference
C. Adjacent interference
D. Disruptive interference

The answer is B. There are many types of harmful interference that can degrade radio transmissions. Other amateur stations could over-modulate and splatter causing interference on adjacent frequencies. Stations using non-directional antennas may cause interference to nearby stations on the same frequency. Interference is influenced by other factors, ex. 1) time of day 2) time of year and season 3) weather conditions 4) sun spot cycle and solar activity 5) height and density of the ionospheric layers.

G1A21 What types of messages may be transmitted by an amateur station to a foreign country for a third party? [97.115, 97.117]
A. Messages for which the amateur operator is paid
B. Messages facilitating the business affairs of any party
C. Messages of a technical nature or remarks of a personal character
D. No messages may be transmitted to foreign countries for third parties

 The answer is C. When transmissions between amateur stations of different countries are permitted, they shall be made in plain language and shall be limited to messages of a technical nature relating to tests and to remarks of a personal character for which, by reason of their unimportance, recourse to the public telecommunications service is not justified.

G1B Antenna structure limitations; good engineering and good amateur practice; beacon operation; restricted operation; retransmitting radio signals

---

G1B01 Up to what height above the ground may you install an antenna structure without needing FCC approval? [97.15a]
A. 50 feet
B. 100 feet
C. 200 feet
D. 300 feet

 The answer is C. If a large antenna or tower is to be installed and the amateur does not have the experience or equipment to do the work, a professional, insured firm should be used. Keep safety in mind when setting up your antenna.

G1B02 If the FCC Rules DO NOT specifically cover a situation, how must you operate your amateur station? [97.101a]
A. In accordance with general licensee operator principles
B. In accordance with good engineering and good amateur practice
C. In accordance with practices adopted by the Institute of Electrical and Electronics Engineers
D. In accordance with procedures set forth by the International Amateur Radio Union

 The answer is B. From the moment you start putting your amateur station together, there will be many situations that are not covered by the FCC rules part 97. Many of these situations will affect the safety of your station. Always seek the help of other experienced amateurs. Use good common safety sense in the installation of your equipment.

G1B03 Which type of station may transmit one-way communications?

[97.203g]
A. Repeater station
B. Beacon station
C. HF station
D. VHF station

The answer is B. The FCC defines a beacon station as an amateur station transmitting communications for the purpose of observation of propagation and reception or other related experimental activities. FCC rules limit automatically controlled beacon station operation to parts of the 28, 50, 144, 222, and 432 MHz amateur bands. Transmitter power is limited to 100 watts and any license class, except Novice can operate a beacon station.

G1B04 Which of the following does NOT need to be true if an amateur station gathers news information for broadcast purposes? [97.113c]
A. The information is more quickly transmitted by amateur radio
B. The information must involve the immediate safety of life of individuals or the immediate protection of property
C. The information must be directly related to the event
D. The information cannot be transmitted by other means

The answer is A. The amateur radio service is ready when emergency information has to be transmitted. During emergency conditions, the amateur service helps by providing communications that can save lives and helps provide information about disasters when conventional methods are disabled. This kind of service to the public is one of the five fundamental purposes for which the amateur rules are designed.

G1B05 Under what limited circumstances may music be transmitted by an amateur station? [97.113e]
A. When it produces no dissonances or spurious emissions
B. When it is used to jam an illegal transmission
C. When it is transmitted on frequencies above 1215 MHz
D. When it is an incidental part of a space shuttle retransmission

The answer is D. The transmission of music in any form is considered an illegal transmission, however, retransmission of space shuttle transmissions is allowed even if music is part of the original transmission.

G1B06 When may an amateur station in two-way communication transmit a message in a secret code in order to obscure the meaning of the communication? [97.113d]
A. When transmitting above 450 MHz
B. During contests
C. Never
D. During a declared communications emergency

The answer is C. The amateur service rules prohibit, under all circumstances, the use of abbreviations, codes or ciphers in all amateur transmissions when they are used to obscure meaning.

G1B07 What are the restrictions on the use of abbreviations or procedural signals in the amateur service? [97.113d]
A. There are no restrictions
B. They may be used if they do not obscure the meaning of a message
C. They are not permitted because they obscure the meaning of a message to FCC monitoring stations
D. Only "10-codes" are permitted

The answer is B. Abbreviations used in amateur transmissions can be used only if they do not obscure the meaning of the message. See question GB106.

G1B08 When are codes or ciphers permitted in two-way domestic amateur communications? [97.113d]
A. Never
B. During contests
C. During nationally declared emergencies
D. On frequencies above 2.3 GHz

The answer is A. See question GB106.

G1B09 When are codes or ciphers permitted in two-way international amateur communications? [97.113d]
A. Never
B. During contests
C. During internationally declared emergencies
D. On frequencies above 2.3 GHz

The answer is A. See question GB106.

G1B10 Which of the following amateur transmissions is NOT prohibited by the FCC Rules? [97.113d]
A. The playing of music
B. The use of obscene or indecent words
C. False or deceptive messages or signals
D. Retransmission of space shuttle communications

The answer is D. See question G1B05.

G1B11 What should you do to keep your station from retransmitting music or signals from a non-amateur station? [97.113d/e]
A. Turn up the volume of your transceiver
B. Speak closer to the microphone to increase your signal strength
C. Turn down the volume of background audio
D. Adjust your transceiver noise blanker

The answer is C. Background noise that may consist of music or transmissions from commercial stations must not be rebroadcast on amateur stations. When transmitting, you must keep the volume of any background audio low in order to stop it from being transmitted by your station.

G1C Transmitter power standards; type acceptance of external RF-power amplifiers; standards for type acceptance of external RF-power amplifiers; HF data emission standards

G1C01 What is the maximum transmitting power an amateur station may use on 3690 kHz? [97.313c1]
A. 200 watts PEP output
B. 1000 watts PEP output
C. 1500 watts PEP output
D. 2000 watts PEP output
The answer is A. This is in the Novice section of the 80-meter wavelength band and limited to 200 watts PEP.

G1C02 What is the maximum transmitting power an amateur station may use on 7080 kHz? [97.313b]
A. 200 watts PEP output
B. 1000 watts PEP output
C. 1500 watts PEP output
D. 2000 watts PEP output
The answer is C. This is in the part of the band reserved for amateurs except Novice and Technician class operators. FCC rules allow 1500 watts.

G1C03 What is the maximum transmitting power an amateur station may use on 10.140 MHz? [97.313c1]
A. 200 watts PEP output
B. 1000 watts PEP output
C. 1500 watts PEP output
D. 2000 watts PEP output
The answer is A. 10.140 MHz is in the new 30-meter wavelength band which extends from 10.1 to 10.15 MHz. This band may be used only by General, Advanced and Extra class operators and is limited to CW, RTTY and data.

G1C04 What is the maximum transmitting power an amateur station may use on 21.150 MHz? [97.313c1]
A. 200 watts PEP output
B. 1000 watts PEP output
C. 1500 watts PEP output
D. 2000 watts PEP output
The answer is A. This is in the Novice section of the 15 meter wavelength

band. 200 watts PEP output is the maximum transmitting power permitted in the 15 meter wavelength Novice sub-band, regardless of the license class of the control operator.

G1C05 What is the maximum transmitting power an amateur station may use on 24.950 MHz? [97.313b]
A. 200 watts PEP output
B. 1000 watts PEP output
C. 1500 watts PEP output
D. 2000 watts PEP output

The answer is C. 24.95 MHz is in the new 12 meter band that extends from 24.89 to 24.99 MHz. 24.95 MHz is in a portion of the band where CW, phone, narrow band FM and SSTV are permitted.

G1C06 External RF power amplifiers designed to operate below what frequency may require FCC type acceptance? [97.315a]
A. 28 MHz
B. 35 MHz
C. 50 MHz
D. 144 MHz

The answer is D. FCC rules (part 97.315) say any external RF power amplifier used with an amateur radio station shall be an accepted type, if it operates below 144 MHz. Approved amplifiers must meet the standards called out (part 97.317) where input and output power limitations, frequency ranges and other rules are listed.

G1C07 Without a grant of FCC type acceptance, how many external RF amplifiers of a given design capable of operation below 144 MHz may you build or modify in one calendar year? [97.315a]
A. None
B. 1
C. 5
D. 10

The answer is B. Under FCC rules (part 97.315) the amateur radio operator may modify or construct an RF power amplifier for use at his station without FCC approval. The amateur operator is limited to producing no more than one unit of one model type per year.

G1C08 Which of the following standards must be met if FCC type acceptance of an external RF amplifier is required? [97.317c6i]
A. The amplifier must not be able to amplify a 28-MHz signal to more than ten times the input power
B. The amplifier must not be capable of reaching its designed output power when driven with less than 50 watts
C. The amplifier must not be able to be operated for more than ten minutes

without a time delay circuit
D. The amplifier must not be able to be modified by an amateur operator

The answer is B. FCC rules (part 97.317) list that full output power must not be reached with less than 50 watts input power.

G1C09 Which of the following would NOT disqualify an external RF power amplifier from being granted FCC type acceptance? [97.317b/c]
A. The capability of being modified by the operator for use outside the amateur bands
B. The capability of achieving full output power when driven with less than 50 watts
C. The capability of achieving full output power on amateur frequencies between 24 and 35 MHz
D. The capability of being switched by the operator to all amateur frequencies below 24 MHz

The answer is D. FCC rules clearly prohibit amplifiers that can be modified for use outside the amateur band, or ones that can output full power with less than 50 watts input and any amplifier that will output full power from 24-35 MHz.

G1C10 What is the maximum symbol rate permitted for packet emissions below 28 MHz? [97.307f3]
A. 300 bauds
B. 1200 bauds
C. 19.6 kilobauds
D. 56 kilobauds

The answer is A. HF packet radio is commonly sent at 300 bauds for frequencies below 28 MHz. Today, most amateur VHF packet radio activity occurs on 2 meters where the most commonly used data rate is 1200 baud. The "baud" is a unit of signaling speed. It is equal to the number of signal events in one second.

G1C11 What is the maximum symbol rate permitted for RTTY emissions below 28 MHz? [97.307f3]
A. 56 kilobauds
B. 19.6 kilobauds
C. 1200 bauds
D. 300 bauds

The answer is D. Radioteletype (RTTY) and data communications are Amateur Radio transmissions that are designed to be received and printed automatically. Often they involve the direct transfer of information (data) between computers. Below 28 MHz, RTTY is commonly sent at 300 baud.

G1D Examination element preparation; examination administration; temporary station identification

G1D01 What telegraphy examination elements may you prepare if you hold a General class license? [97.507a2]
A. None
B. Element 1A only
C. Element 1B only
D. Elements 1A and 1B

The answer is B. The Novice code test element 1A requires testing at five words per minute. FCC rules (part 97.507) call out the requirements for examination elements.

G1D02 What written examination elements may you prepare if you hold a General class license? [97.507a2&3]
A. None
B. Element 2 only
C. Elements 2 and 3A
D. Elements 2, 3A and 3B

The answer is C. Element 2: Basic law comprising rules and regulations essential to the beginner's operation, including sufficient elementary radio theory. Element 3A: Beginner level theory and regulations are called out in FCC rules part 97.507.

G1D03 What license examinations may you administer if you hold a General class license? [97.511b1]
A. None
B. Novice only
C. Novice and Technician
D. Novice, Technician and General

The answer is C. The Novice class test consists of element 2 and element 1A. The type of exam is a 30-question written examination and a 5 word-per-minute code test. The Technician class test consists of elements 2 and 3A. The type of exam is a 55-question written examination (in two parts: 30 questions from element 2, 25 questions from element 3A) with no Morse code requirement. FCC rules part 97.511b examination requirements.

G1D04 What minimum examination elements must an applicant pass for a Novice license? [97.501e]
A. Element 2 only
B. Elements 1A and 2
C. Elements 2 and 3A
D. Elements 1A, 2 and 3A

The answer is B. See question G1D03 and FCC rules 97.501e.

G1D05 What minimum examination elements must an applicant pass for a Technician license? [97.501d]
A. Element 2 only
B. Elements 1A and 2
C. Elements 2 and 3A
D. Elements 1A, 2 and 3A

The answer is C. See question G1D03 and FCC rules part 97.501d.

G1D06 What minimum examination elements must an applicant pass for a Technician license with HF privileges? [97.301e/501d]
A. Element 2 only
B. Elements 1A and 2
C. Elements 2 and 3A
D. Elements 1A, 2 and 3A

The answer is D. The Technician Plus class (HF privileges) license requires elements 2 and 3A like the Technician class, and element 1A, the 5 word-per-minute code test from the Novice test.

G1D07 What are the requirements for administering Novice examinations? [97.511a/b]
A. Three VEC-accredited General class or higher VEs must be present
B. Two VEC-accredited General class or higher VEs must be present
C. Two General class or higher VEs must be present, but only one need be VEC accredited
D. Any two General class or higher VEs must be present

The answer is A. For all licenses including Novice operator, the examinations are administered by three local amateur operators certified as Volunteer Examiners (VE's). VE teams usually provide the information on local VHF networks as to when and where examination sessions are to be held. Their efforts are coordinated by a VEC (Volunteer-Examiner Coordinator) who accredits them to serve as a VE.

G1D08 When may you participate as an administering Volunteer Examiner (VE) for a Novice license examination? [97.507a]
A. Once you have notified the FCC that you want to give an examination
B. Once you have a Certificate of Successful Completion of Examination (CSCE) for General class
C. Once you have prepared telegraphy and written qualified examinations for the Novice license, or obtained them from a qualified supplier
D. Once you have received both your FCC-issued General class or higher license in the mail and VEC accreditation

The answer is D. See question G1D07

G1D09 If you are a Technician licensee with a Certificate of Successful Completion of Examination (CSCE) for General privileges, how do you

identify your station when transmitting on 14.035 MHz? [97.119e2]
A. You must give your call sign and the location of the VE examination where you obtained the CSCE
B. You must give your call sign, followed by the slant mark "/", followed by the identifier "AG"
C. You may not operate on 14.035 MHz until your new license arrives
D. No special form of identification is needed

The answer is B. When you upgrade from Novice to Technician or Technician to General class you will receive a Certificate of Successful Completion of Examination (CSCE) showing the exam elements you passed. The CSCE authorizes you to operate with your new privileges. When you use these new privileges, you must use your call sign, followed by the slant mark ("/": on voice say "stroke or slant") and the letters KT for Novices upgrading to Technician. For Technician's upgrading to General class use the letters AG.

G1D10 If you are a Technician licensee with a Certificate of Successful Completion of Examination (CSCE) for General privileges, how do you identify your station when transmitting phone emissions on 14.325 MHz? [97.119e2]
A. No special form of identification is needed
B. You may not operate on 14.325 MHz until your new license arrives
C. You must give your call sign, followed by any suitable word that denotes the slant mark and the identifier "AG"
D. You must give your call sign and the location of the VE examination where you obtained the CSCE

The answer is C. See question G1D09

G1D11 If you are a Technician licensee with a Certificate of Successful Completion of Examination (CSCE) for General privileges, when must you add the special identifier "AG" after your call sign? [97.119e2]
A. Whenever you operate using your new frequency privileges
B. Whenever you operate
C. Whenever you operate using Technician frequency privileges
D. A special identifier is not required as long as your General class license application has been filed with the FCC

The answer is A. See question G1D09

SUBELEMENT G2
OPERATING PROCEDURES
[3 exam questions - 3 groups]

G2A Phone, RTTY, repeater, VOX and full break-in CW

G2A01 Which sideband is commonly used for 20-meter phone operation?
A. Upper
B. Lower
C. Amplitude compandored
D. Double

The answer is A. In the system of amplitude modulation, both the upper and lower sidebands are produced and transmitted. In the single-sideband system, both the upper and lower sidebands are produced. However, either the upper or lower sidebands are suppressed in the transmitter and the other is transmitted, hence the term single-sideband transmission.

The upper sidebands are commonly used in the 20, 15 and 10 meter bands. The lower sidebands are used in the 40, 80 and 160 meter bands.

G2A02 Which sideband is commonly used on 3925-kHz for phone operation?
A. Upper
B. Lower
C. Amplitude compandored
D. Double

The answer is B. See answer G2A01.

G2A03 In what segment of the 80-meter band do most RTTY transmissions take place?
A. 3610 - 3630 kHz
B. 3500 - 3525 kHz
C. 3700 - 3750 kHz
D. 3775 - 3825 kHz

The answer is A. On the 80-meter band RTTY is legally permitted between 3.5 and 3.75 MHz.

G2A04 In what segment of the 20-meter band do most RTTY transmissions take place?
A. 14.000 - 14.050 MHz
B. 14.075 - 14.100 MHz
C. 14.150 - 14.225 MHz
D. 14.275 - 14.350 MHz

The answer is B. RTTY is legally permitted between 14.000 and 14.150 MHz; however, by agreement among amateurs, it is confined to 14.075 to 14.100 MHz.

G2A05 What is the Baudot code?
A. A 7-bit code, with start, stop and parity bits
B. A 7-bit code in which each character has four mark and three space bits
C. A 5-bit code, with additional start and stop bits
D. A 6-bit code, with additional start, stop and parity bits

The answer is C. Baudot is a code used in radioteletype. In the Baudot code, each character is made up of five data elements plus a start element and a stop element. An element is either a space(0) or a mark(1). The different characters (letters of the alphabet, numbers, etc.) have different combinations of spaces and marks. Figure G2A05-A shows the letter "J". The five data pulses that distinguish the letter J are: mark, mark, space, mark, space. Figure G2A05-B illustrates the letter "Y". The five data pulses that distinguish the letter Y are: mark, space, mark, space, mark. There is a start pulse in front of the five data pulses. This start pulse is always a space pulse. After the five data pulses, there is a stop pulse. This stop pulse is always a mark pulse. When the speed is 60 words per minute, the time duration for each data pulse and start pulse is 22 ms(milliseconds); the duration for the stop pulse is 31 ms. Higher code speeds have shorter duration pulses.

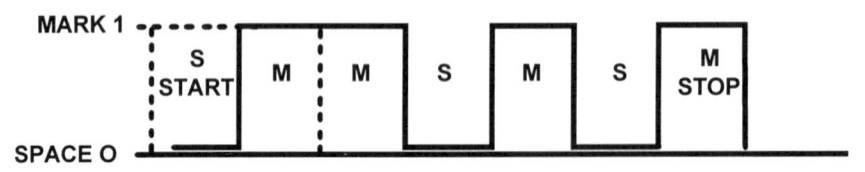

Fig. G2A05-A The letter "J" in Baudot code. The "S" stands for space and the "M" for mark.

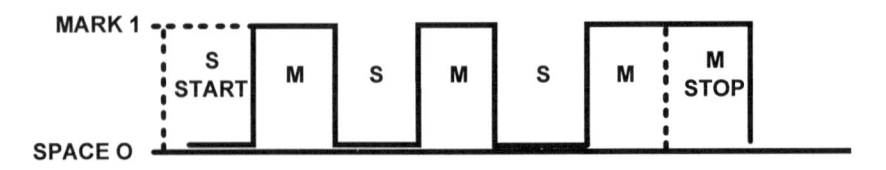

Fig. G2A05-B The letter "Y" in Baudot code. The "S" stands for space and the "M" for mark.

G2A06 What is ASCII?
A. A 7-bit code, with additional start, stop and parity bits
B. A 7-bit code in which each character has four mark and three space bits
C. A 5-bit code, with additional start and stop bits

D. A 5-bit code in which each character has three mark and two space bits

The answer is A. The term ASCII (pronounced "askee") is the abbreviation for American Standard Code for Information Interchange. It was developed in 1968 for computer and communications uses. Whereas the Baudot Code contains five data bits, ASCII has seven bits. The seven data bits permit many more combinations (128) for the ASCII Code than the five data bits of the Baudot Code. ASCII also provides both upper and lower case letters, a more complete set of punctuation and a number of control characters. HF ASCII transmissions are normally at 110 or 300 baud using a 170 Hz frequency shift. Higher data rates are used on the 10-meter band where the transmission speed of 1200 baud is permitted. Up to 19,600 baud transmission speed is allowed on 222 MHz and above. An eighth bit may be added to the ASCII Code for error detection. It is called a "parity" bit.

G2A07 What is the most common frequency shift for RTTY emissions in the amateur HF bands?
A. 85 Hz
B. 170 Hz
C. 425 Hz
D. 850 Hz

The answer is B. Most modems used for amateur RTTY have shift capabilities of both 850 Hz and 170 Hz. 850 Hz was originally used. However, today 170 Hz is the normal shift for RTTY communications.

G2A08 What are the two major AMTOR operating modes?
A. Mode AM and Mode TR
B. Mode A (ARQ) and Mode B (FEC)
C. Mode C (CRQ) and Mode D (DEC)
D. Mode SELCAL and Mode LISTEN

The answer is B. Amateur Teleprinting Over Radio (AMTOR) is a more reliable transmission system than Baudot RTTY. AMTOR is a synchronous system with the ability to detect errors. There are a few modes of operation. Mode A (ARQ) stands for Automatic Repeat Request and Mode B (FEC) stands for Forward Error Corrections. In the ARQ mode, the two stations continuously check each other and the signal is repeated only when requested. In the FEC mode, each character is sent twice. Mode L has been added by amateurs for monitoring AMTOR transmissions.

G2A09 What is the usual input/output frequency separation for a 10-meter station in repeater operation?
A. 100 kHz
B. 600 kHz
C. 1.6 MHz
D. 170 Hz

The answer is A. The separation of 100 KHz is in general use on 10 meters. This separation is not specified in the FCC rules, but is agreed upon by the local users of the repeater. Typical input/output offsets are: for 2 meters = 600 KHz, 1.25 meters = 1.6 MHz, 70 cm = 5 MHz.

G2A10 What is the circuit called which causes a transmitter to automatically transmit when an operator speaks into its microphone?
A. VXO
B. VOX
C. VCO
D. VFO

The answer is B. VOX stands for VOICE OPERATED TRANSMISSION. The VOX circuit in a system controls the transmit-receive changeover circuit. When the operator speaks into the microphone, the transmitter is automatically switched "on" and the receiver "off". When the operator ceases talking, the transmitter is automatically switched "off" and the receiver is turned "on". Modern transceivers have the VOX switch built in. When the operator speaks into the microphone, the transceiver is automatically switched to transmit, then back to receive when the operator ceases talking. VOX can be very handy because you can listen during pauses and keep both hands free.

G2A11 Which of the following describes full break-in telegraphy?
A Breaking stations send the Morse code prosign BK
B. Automatic keyers are used to send Morse code instead of hand keys
C. An operator must activate a manual send/receive switch before and after every transmission
D. Incoming signals are received between transmitted key pulses

The answer is D. Full break-in telegraphy, abbreviated "QSK" allows an operator who is transmitting to hear the other operator between his sending of dots and dashes. This allows for more efficient operating. This feature is most useful to the CW operator who is able to send code quickly.

G2B Operating courtesy, antenna orientation and HF operations, including logging practices; ITU Regions

---

G2B01 If you are the net control station of a daily HF net, what should you do if the frequency on which you normally meet is in use just before the net begins?
A. Reduce your output power and start the net as usual
B. Increase your power output so that net participants will be able to hear you over the existing activity
C. Cancel the net for that day
D. Conduct the net on a frequency 3 to 5 kHz away from the regular net frequency

The answer is D. A net is a group of amateurs who meet on the air to pass traffic or to communicate with each other on a specific topic. One station (called the net control station) usually directs the net. Most nets meet on a regular basis. CW traffic nets are a good opportunity to improve your code speed and serve as a training ground for amateurs who desire to enhance their on-air operating skills. Slow-speed CW nets are very popular with new amateurs.

G2B02 If a net is about to begin on a frequency which you and another station are using, what should you do?
A. As a courtesy to the net, move to a different frequency
B. Increase your power output to ensure that all net participants can hear you
C. Transmit as long as possible on the frequency so that no other stations may use it
D. Turn off your radio

The answer is A. Every day courtesy works very well in these type of situations. If the net control station found the frequency was being used by you first, he would move the net to a different frequency, as a courtesy to you.

G2B03 If propagation changes during your contact and you notice increasing interference from other activity on the same frequency, what should you do?
A. Tell the interfering stations to change frequency, since you were there first
B. Report the interference to your local Amateur Auxiliary Coordinator
C. Turn on your amplifier to overcome the interference
D. Move your contact to another frequency

The answer is D. Another situation where giving a little by changing frequency will make things better for everyone on the air. By showing operating courtesy you will invite others to do the same.

G2B04 When selecting a CW transmitting frequency, what minimum frequency separation from a contact in progress should you allow to minimize interference?
A. 5 to 50 Hz
B. 150 to 500 Hz
C. 1 to 3 kHz
D. 3 to 6 kHz

The answer is B. Even though CW is merely an interruption of a single frequency carrier, a CW signal does occupy a certain amount of frequency spectrum. The dit-dah interruption of a carrier produces a limited degree of modulation, which results in a limited bandwidth. This bandwidth is determined by the speed of sending and the shape of the keyed signal. The bandwidth required for a CW signal is much less than that of a phone signal. A few hundred Hz will provide the separation required.

G2B05 When selecting a single-sideband phone transmitting frequency, what minimum frequency separation from a contact in progress should you allow (between suppressed carriers) to minimize interference?
A. 150 to 500 Hz
B. Approximately 3 kHz
C. Approximately 6 kHz
D. Approximately 10 kHz

The answer is B. The audio of a phone signal is limited to between 2.5 and 3.0 kHz. The minimum separation that can be used without causing interference would be 3.0 kHz from any adjacent signals. If you can hear someone on SSB, move at least 3 kHz away.

G2B06 When selecting an RTTY transmitting frequency, what minimum frequency separation from a contact in progress should you allow (center-to-center) to minimize interference?
A. 60 Hz
B. 250 to 500 Hz
C. Approximately 3 kHz
D. Approximately 6 kHz

The answer is B. When RTTY is used, the transmitter carrier is shifted between two frequencies. The shift or frequency difference between the two frequencies is commonly 170 Hz. Thus, we must allow a bandwidth of at least this amount. To stay away from adjacent signals we must allow a few hundred Hertz separation. 500 Hz will work well for both CW and RTTY.

G2B07 What is an azimuthal map?
A. A map projection centered on the North Pole
B. A map projection centered on a particular location, used to determine the shortest path between points on the surface of the earth
C. A map that shows the angle at which an amateur satellite crosses the equator
D. A map that shows the number of degrees longitude that an amateur satellite appears to move westward at the equator with each orbit

The answer is B. It is the best kind of map to use when trying to orient a directional antenna for finding the shortest path to a distant station. Long range transmissions do not go in a straight line. When we aim our signals around the world we use an azimuthal map that will take into account the curvature of the earth.

G2B08 What is the most useful type of map to use when orienting a directional HF antenna toward a distant station?

A. Azimuthal
B. Mercator
C. Polar projection
D. Topographical

The answer is A. Azimuthal maps will show the shortest distance between two locations and make it easy to orient the antenna. See question G2B07.

G2B09 A directional antenna pointed in the long-path direction to another station is generally oriented how many degrees from its short-path heading?
A. 45 degrees
B. 90 degrees
C. 180 degrees
D. 270 degrees

The answer is C. The long path is 180 degrees opposite to the short path in a directional antenna. The antenna would have to be pointed 180 degrees away from the short-path.

G2B10 What is a band plan?
A. A guideline for using different operating modes within an amateur band
B. A guideline for deviating from FCC amateur frequency band allocations
C. A plan of operating schedules within an amateur band published by the FCC
D. A plan devised by a club to best use a frequency band during a contest

The answer is A. A band plan is a plan for allocating frequency segments in a band for specific types of operating. This allows for orderly operating practice with a minimum of interference. Band plans are developed by amateur organizations and are not part of the FCC rules. However, the FCC does encourage band plans and considers them to be in accordance with good amateur practice. Amateur operators have put band plans into use by reserving segments of world wide bands for working foreign stations, slow-scan television and satellite reception.

G2B11 In which International Telecommunication Union Region is the continental United States?
A. Region 1
B. Region 2
C. Region 3
D. Region 4

The answer is B. The International Telecommunication Union (I.T.U.) is the world organization that allocates frequency use to the various services. The I.T.U. has, for purposes of frequency allocation, divided the world into three regions. Region 1 consists principally of Europe and Africa. Region 2 includes North America and South America. Region 3 consists primarily of Southern Asia and Australia.

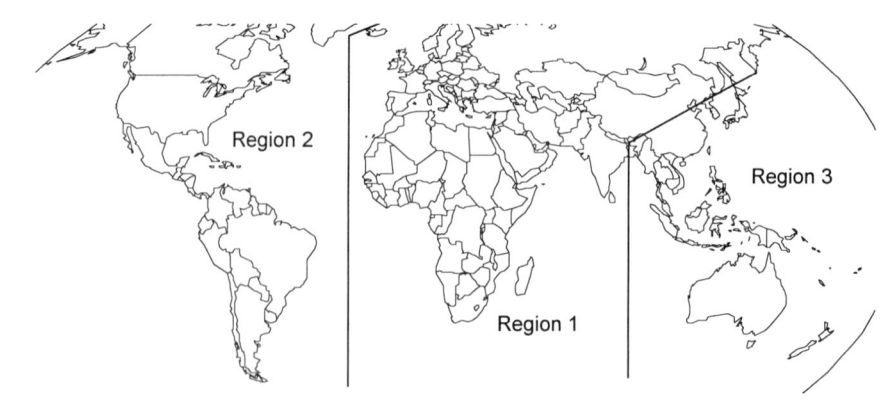

Figure G2B11. ITU Regions

> **G2C** Emergencies, including drills, communications and amateur auxiliary to FOB

G2C01 What means may an amateur station in distress use to attract attention, make known its condition and location, and obtain assistance?
A. Only Morse code signals sent on internationally recognized emergency channels
B. Any means of radiocommunication, but only on internationally recognized emergency channels
C. Any means of radiocommunication
D. Only those means of radiocommunication for which the station is licensed

The answer is C. When life or property is in immediate danger the normal FCC restrictions are suspended. You can request immediate emergency help by transmitting on whatever frequency offers you the best chance of getting help.

G2C02 During a disaster in the US, when may an amateur station make transmissions necessary to meet essential communication needs and assist relief operations?
A. When normal communication systems are overloaded, damaged or disrupted
B. Only when the local RACES net is activated
C. Never; only official emergency stations may transmit in a disaster
D. When normal communication systems are working but are not convenient

The answer is A. The position of the FCC is very clear on emergency communications. They not only allow, but encourage licensed amateur operators to assist in emergencies. FCC rules state: When normal

OPERATING PROCEDURES 29

communication systems are overloaded, damaged or disrupted because a disaster has occurred, an amateur station may make any transmission necessary to meet essential communication needs and help relief efforts. During major disasters, the FCC may suspend its rules to help the immediate problem. This is called a temporary state of communication emergency. When the FCC declares a state of emergency, they will issue any special conditions and rules that should be observed by amateur stations during the emergency.

G2C03 If a disaster disrupts normal communications in your area, what may the FCC do?
A. Declare a temporary state of communication emergency
B. Temporarily seize your equipment for use in disaster communications
C. Order all stations across the country to stop transmitting at once
D. Nothing until the President declares the area a disaster area
   The answer is A. See answer to question G2C03.

G2C04 If a disaster disrupts normal communications in an area, what would the FCC include in any notice of a temporary state of communication emergency?
A. Any additional test questions needed for the licensing of amateur emergency communications workers
B. A list of organizations authorized to temporarily seize your equipment for disaster communications
C. Any special conditions requiring the use of non-commercial power systems
D. Any special conditions and special rules to be observed by stations during the emergency
   The answer is D. See answer to question G2C03.

G2C05 During an emergency, what power output limitations must be observed by a station in distress?
A. 200 watts PEP
B. 1500 watts PEP
C. 1000 watts PEP during daylight hours, reduced to 200 watts PEP during the night
D. There are no limitations during an emergency
   The answer is D. The key word is emergency. During an emergency, the FCC rules and regulations are suspended. See answer to question C2C01.

G2C06 During a disaster in the US, what frequencies may be used to obtain assistance?
A. Only frequencies in the 80-meter band
B. Only frequencies in the 40-meter band
C. Any frequency

D. Any United Nations approved frequency

The answer is C. The key word is disaster. During a disaster, emergency conditions exist and FCC rules are suspended.

G2C07 If you are communicating with another amateur station and hear a station in distress break in, what should you do?
A. Continue your communication because you were on frequency first
B. Acknowledge the station in distress and determine its location and what assistance may be needed
C. Change to a different frequency so the station in distress may have a clear channel to call for assistance
D. Immediately cease all transmissions because stations in distress have emergency rights to the frequency

The answer is B. The station making the emergency transmission will be prepared to give information on its location and what assistance is needed. You must get that information and send it to the proper authorities so they can provide the needed help.

G2C08 Why do stations in the Radio Amateur Civil Emergency Service (RACES) participate in training tests and drills?
A. To practice orderly and efficient operations for the civil defense organization they serve
B. To ensure that members attend monthly on-the-air meetings
C. To ensure that RACES members are able to conduct tests and drills
D. To acquaint members of RACES with other members they may meet in an emergency

The answer is A. The only way to insure that emergency plans for radio amateurs assisting in emergency services are going to operate smoothly during a real disaster is to practice. The training tests and drills are designed to sharpen the skills of the participants.

G2C09 What type of messages may be transmitted to an amateur station in a foreign country?
A. Messages of any type
B. Messages that are not religious, political, or patriotic in nature
C. Messages of a technical nature or personal remarks of relative unimportance
D. Messages of any type, but only if the foreign country has a third-party communications agreement with the US

The answer is C. Permitted transmissions should not include business messages or political discussions. Talking to Hams in other countries is fun, so keep your communications to a technical or personal nature.

G2C10 What is the Amateur Auxiliary to the FCC's Field Operations Bureau?

A. Amateur volunteers who are formally enlisted to monitor the airwaves for rules violations
B. Amateur volunteers who conduct amateur licensing examinations
C. Amateur volunteers who conduct frequency coordination for amateur VHF repeaters
D. Amateur volunteers who use their station equipment to help civil defense organizations in times of emergency

The answer is A. The Communication Amendments Act of 1982 (Public Law 97-259) authorizes the FCC to use amateur volunteers to help monitor the airwaves for rules violations. The FCC's Field Operations Bureau has created an Amateur Auxiliary which is administered by the ARRL. This allows the amateur radio operators to assist in policing their own airwaves.

G2C11 What are the objectives of the Amateur Auxiliary to the FCC's Field Operations Bureau?
A. To conduct efficient and orderly amateur licensing examinations
B. To encourage amateur self-regulation and compliance with the rules
C. To coordinate repeaters for efficient and orderly spectrum usage
D. To provide emergency and public safety communications

The answer is B. The primary purpose of the Amateur Auxiliary is to provide help to other amateurs and to find the causes of amateur problems. It is NOT to enforce the rules. Ham operators want self-regulation and pride themselves on how successful they have been.

SUBELEMENT G3
RADIO WAVE PROPAGATION
[3 exam questions - 3 groups]

G3A Ionospheric disturbances; sunspots and solar radiation

G3A01 What can be done at an amateur station to continue communications during a sudden ionospheric disturbance?
A. Try a higher frequency
B. Try the other sideband
C. Try a different antenna polarization
D. Try a different frequency shift
  The answer is A. Since different frequencies are affected differently by the ionosphere, the only thing that you can do is to try a different band, preferably a higher frequency band. Sudden ionospheric disturbances affect the lower frequencies more than the higher frequencies because the lower frequencies are absorbed by the D layer.

G3A02 What effect does a sudden ionospheric disturbance have on the daylight ionospheric propagation of HF radio waves?
A. It disrupts higher-latitude paths more than lower-latitude paths
B. It disrupts signals on lower frequencies more than those on higher frequencies
C. It disrupts communications via satellite more than direct communications
D. None, only areas on the night side of the earth are affected
  The answer is B. The ionospheric disturbance will affect the lower frequencies such as 160, 80 and 40 meters. During severe disturbances even 20 meters. These disturbances increase the absorption of the lower ionospheric levels.

G3A03 How long does it take the increased ultraviolet and X-ray radiation from solar flares to affect radio-wave propagation on the earth?
A. The effect is instantaneous
B. 1.5 seconds
C. 8 minutes
D. 20 to 40 hours
  The answer is C. It takes about 8 minutes for the effects of ultraviolet radiation to show up as a disturbance of radio waves. This is because ultraviolet radiation travels at about the speed of light, 186,000 miles per second, and the sun is 92,900,000 miles away from the earth.

G3A04 What is solar flux?

RADIO WAVE PROPAGATION

A. The density of the sun's magnetic field
B. The radio energy emitted by the sun
C. The number of sunspots on the side of the sun facing the earth
D. A measure of the tilt of the earth's ionosphere on the side toward the sun

The answer is B. The sun emits ultraviolet rays and particles that cause ionization of the ionosphere. We refer to the sun's radiated energy as solar flux. The solar flux or radiated energy emitted by the sun can be measured with special receiving equipment.

G3A05 What is the solar-flux index?
A. A measure of solar activity that is taken annually
B. A measure of solar activity that compares daily readings with results from the last six months
C. Another name for the American sunspot number
D. A measure of solar activity that is taken at a specific frequency

The answer is D. The radiation from the sun can be picked up as noise on a receiver. The amount of noise picked up depends upon the amount of solar activity and varies a great deal. The unit used to measure and compare solar radiation is called the "solar flux index". The solar flux index is measured daily at 1700 (5:00 PM) UTC at an observatory in Ottawa, Canada. This measurement is made on a frequency of 2800 MHz. The readings are broadcast by the National Bureau of Standards radio station WWV and WWVH. You can also tune into the radio wave propagation forecast at 18 minutes past the hour on 5, 10, 15 and 20 MHz, AM or USB. The reports are 45-seconds long and will give you the latest ionospheric conditions.

G3A06 What is a geomagnetic disturbance?
A. A sudden drop in the solar-flux index
B. A shifting of the earth's magnetic pole
C. Ripples in the ionosphere
D. A dramatic change in the earth's magnetic field over a short period of time

The answer is D. The earth itself is a magnet with magnetic poles and a magnetic field. The earth's magnetic field affects radio wave propagation. Strong radiation of particles from the sun during periods of intense solar activity will cause disturbances in the earth's magnetic field, which in turn, will affect radio wave propagation. These disturbances in the earth's magnetic field are called "geomagnetic disturbances".

G3A07 At which latitudes are propagation paths more sensitive to geomagnetic disturbances?
A. Those greater than 45 degrees latitude
B. Those between 5 and 45 degrees latitude
C. Those near the equator

D. All paths are affected equally

The answer is A. The magnetic disturbances are more intense north of latitude 45. Geomagnetic disturbances are associated with the Aurora Borealis or Northern Lights.

G3A08 What can be the effect of a major geomagnetic storm on radio-wave propagation?
A. Improved high-latitude HF propagation
B. Degraded high-latitude HF propagation
C. Improved ground-wave propagation
D. Improved chances of UHF ducting

The answer is B. Ionospheric ionization during radio storms will degrade worldwide HF communications in northern latitudes above 45 degrees.

G3A09 What influences all radio communication beyond ground-wave or line-of-sight ranges?
A. Solar activity
B. Lunar tidal effects
C. The F1 region of the ionosphere
D. The F2 region of the ionosphere

The answer is A. Solar activity has a major influence on radio wave propagation because the ionosphere becomes highly ionized and causes increased radio wave absorption.

G3A10 Which two types of radiation from the sun influence propagation?
A. Subaudible- and audio-frequency emissions
B. Electromagnetic and particle emissions
C. Polar-region and equatorial emissions
D. Infrared and gamma-ray emissions

The answer is B. Electromagnetic radiation affects the earth's magnetic field and changes radio wave propagation. Particle emissions will affect the ionospheric conditions and radio wave absorption.

G3A11 When sunspot numbers are high, how is the ionosphere affected?
A. High-frequency radio signals are absorbed
B. Frequencies up to 100 MHz or higher are normally usable for long-distance communication
C. Frequencies up to 40 MHz or higher are normally usable for long-distance communication
D. High-frequency radio signals become weak and distorted

The answer is C. As sunspot activity increases, the ionosphere becomes more and more ionized, resulting in increased radio wave absorption. The low frequencies will be affected more than high frequencies. As the sunspot

RADIO WAVE PROPAGATION   35

activity increases, the interference will be found on higher and higher bands.

> **G3B** Maximum usable frequency, propagation "hops"

G3B01 If the maximum usable frequency on the path from Minnesota to France is 22 MHz, which band should offer the best chance for a successful contact?
A. 10 meters
B. 15 meters
C. 20 meters
D. 40 Meters

The answer is B. The MAXIMUM USABLE FREQUENCY (MUF) is the highest frequency that can be used for communication between two specific areas. As the frequency of the signal is increased, a point is reached where the signal penetrates the ionosphere and does not return to earth. The best chance for a successful QSO would be to use the next lowest amateur frequency band from the MUF. In this question, it is the 15 meter wavelength band (21.0 to 21.45 MHz)

G3B02 If the maximum usable frequency on the path from Ohio to Germany is 17 MHz, which band should offer the best chance for a successful contact?
A. 80 meters
B. 40 meters
C. 20 meters
D. 2 meters

The answer is C. If the MUF IS 17 MHz, the next lower amateur band would be 20 meters (14.0 to 14.35 MHz). See question G3B01.

G3B03 If the maximum usable frequency (MUF) is high and HF radio-wave propagation is generally good for several days, a similar condition can usually be expected how many days later?
A. 7
B. 14
C. 28
D. 90

The answer is C. The MUF and HF radiocommunications are due to solar activity which correlates with sunspots on the sun's surface. Since the sun's rotation cycle is approximately 28 days, it will take that amount of time for the same sunspots to reappear with its consequent similar propagation conditions.

G3B04 What is one way to determine if the maximum usable frequency (MUF) is high enough to support 28-MHz propagation between your

station and western Europe?
A. Listen for signals on the 10-meter beacon frequency
B. Listen for signals on the 20-meter beacon frequency
C. Listen for signals on the 39-meter broadcast frequency
D. Listen for WWVH time signals on 20 MHz

The answer is A. 28 MHz is on the 10-meter band, so listen on ten meters for the radio wave propagational forecast transmitted at 18 minutes after the hour.

G3B05 What usually happens to radio waves with frequencies below the maximum usable frequency (MUF) when they are sent into the ionosphere?
A. They are bent back to the earth
B. They pass through the ionosphere
C. They are completely absorbed by the ionosphere
D. They are changed to a frequency above the MUF

The answer is A. The frequency above which radio waves penetrate the ionosphere and do not return to earth is called the MUF. Frequencies below MUF are bent by the ionosphere and returned to earth. The MUF between two points depends on solar conditions and the time of day. Ionization of the upper atmosphere is worst during the day and summer. See figure G3B05.

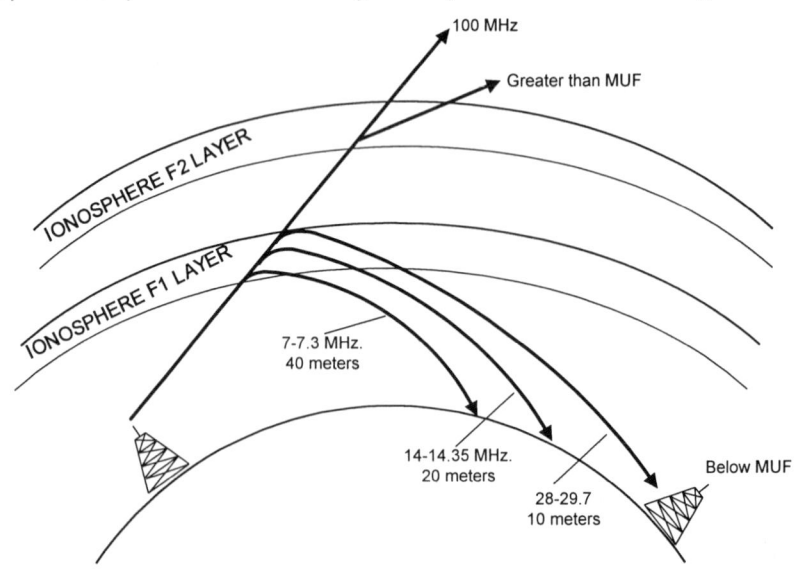

Figure G3B05

G3B06 Where would you tune to hear beacons that would help you determine propagation conditions on the 20-meter band?
A. 28.2 MHz

B. 21.1 MHz
C. 14.1 MHz
D. 14.2 MHz

The answer is C. 14.1 is on the 20-meter wave length band. See question G3A05.

G3B07 During periods of low solar activity, which frequencies are the least reliable for long-distance communication?
A. Frequencies below 3.5 MHz
B. Frequencies near 3.5 MHz
C. Frequencies on or above 10 MHz
D. Frequencies above 20 MHz

The answer is D. As the frequency of the signal is increased, a point is reached where the signal will penetrate the ionosphere and not be returned to earth. See question G3B01.

G3B08 At what point in the solar cycle does the 20-meter band usually support worldwide propagation during daylight hours?
A. At the summer solstice
B. Only at the maximum point of the solar cycle
C. Only at the minimum point of the solar cycle
D. At any point in the solar cycle

The answer is D. The 20-meter wavelength band (14.00 to 14.35 MHz) is low enough in frequency to provide good world-wide communication. Communication on the 20-meter band is not as dependent on solar conditions because of its lower frequency. During the day you can hear signals from Europe, Africa and South America. In the early afternoon and evening, 20 meters will provide good, long distance contacts to everywhere in the Pacific and Japan.

G3B09 What is one characteristic of gray-line propagation?
A. It is very efficient
B. It improves local communications
C. It is very poor
D. It increases D-region absorption

The answer is A. The gray line is a transition area between daylight and darkness. One side of the earth is coming into sunrise, and the other side is just past sunset. Radio wave propagation along the gray line is very efficient because the D layer, which absorbs HF signals, disappears on the sunset side of the gray line. On the sunrise side, the D layer has not had time to build up.

G3B10 What is the maximum distance along the earth's surface that is normally covered in one hop using the F2 region?
A. 180 miles

B. 1200 miles
C. 2500 miles
D. None; the F2 region does not support radio-wave propagation

The answer is C. The F2 region is the highest layer in the ionosphere, thus we can get the greatest distance along the earth's surface in one hop. During the day the average height of the F2 layer is 225 miles above the earth. See figure G3B10.

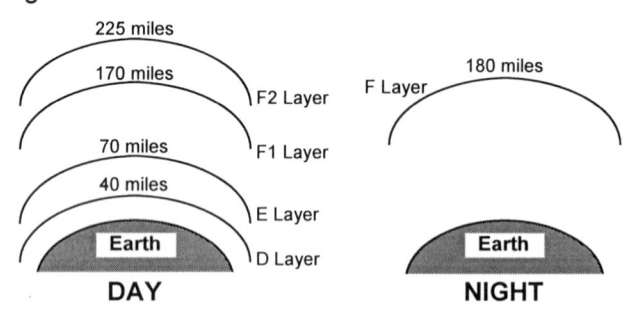

Figure G3B10

G3B11 What is the maximum distance along the earth's surface that is normally covered in one hop using the E region?
A. 180 miles
B. 1200 miles
C. 2500 miles
D. None; the E region does not support radio-wave propagation

The answer is B. The E layer is lower than the F2 layer. During the day the average height of the E layer is 70 miles above the earth so we only get a 1200 mile hop.

> **G3C** Height of ionospheric regions, critical angle and frequency, HF scatter

G3C01 What is the average height of maximum ionization of the E region?
A. 45 miles
B. 70 miles
C. 200 miles
D. 1200 miles

The answer is B. The average height is 70 miles above the earth during the day. See figure G3B10.

G3C02 When can the F2 region be expected to reach its maximum height at your location?

A. At noon during the summer
B. At midnight during the summer
C. At dusk in the spring and fall
D. At noon during the winter

The answer is A. The maximum ionization of the ionosphere will occur during the day in the summer. This will cause the F2 layer to rise to its maximum height. During the evening, the D and E layers disappear and the F2 and F1 layers combine to form a single F layer. The F layer can rise to 180 miles at night.

G3C03 Why is the F2 region mainly responsible for the longest-distance radio-wave propagation?
A. Because it exists only at night
B. Because it is the lowest ionospheric region
C. Because it is the highest ionospheric region
D. Because it does not absorb radio waves as much as other ionospheric regions

The answer is C. The F2 layer is the highest ionospheric level during the day. See figure G3B10.

G3C04 What is the "critical angle" as used in radio-wave propagation?
A. The lowest takeoff angle that will return a radio wave to the earth under specific ionospheric conditions
B. The compass direction of a distant station
C. The compass direction opposite that of a distant station
D. The highest takeoff angle that will return a radio wave to the earth under specific ionospheric conditions

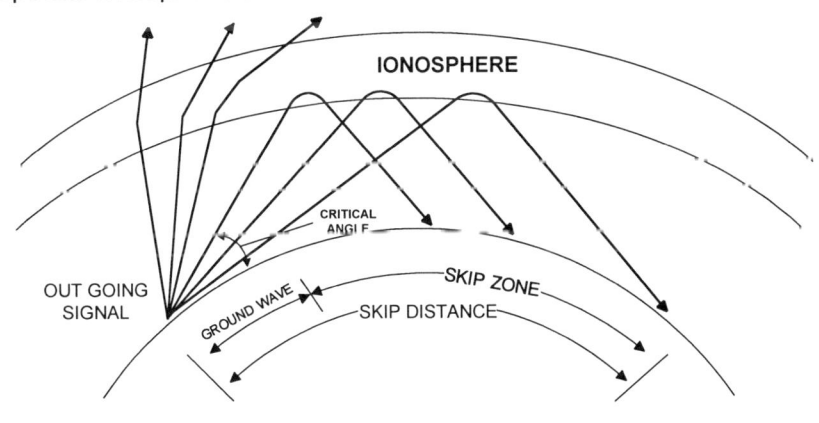

Figure G3C04

The answer is D. The angle of radiation is the angle between the sky wave and the earth's surface. This is shown in Figure G3C04. It can be seen that as the angle of radiation is reduced, the skywave is refracted back to a more distant point on earth. However, as the angle of radiation is increased, the sky wave penetrates the ionosphere and is NOT returned to earth. The angle above which there is no return of the signal to the earth and below which signals ARE returned to earth, is called the "critical angle". The value of the critical angle depends on the frequency and other ionospheric conditions.

G3C05 What is the main reason the 160-, 80- and 40-meter amateur bands tend to be useful only for short-distance communications during daylight hours?
A. Because of a lack of activity
B. Because of auroral propagation
C. Because of D-region absorption
D. Because of magnetic flux

The answer is C. The main reason is absorption by the D layer which limits the range of communications for these bands during the day. The D layer is more of an absorbing layer rather than a reflecting layer.

G3C06 What is a characteristic of HF scatter signals?
A. High intelligibility
B. A wavering sound
C. Reversed modulation
D. Reversed sidebands

The answer is B. See figure G3C06. Note the "skip zone". It is the area where there is ordinarily no reception. The sky wave skips over it. A scatter mode of propagation is used to "penetrate" the skip zone. Scatter refers to the random refraction or reflection of radio waves by irregularities in the earth's atmosphere or the earth's surface. Scatter signals are invariably weak. However, scatter signals can provide reliable communications on some VHF frequencies up to several hundred miles if high power transmitters, efficient antennas and sensitive receivers are used.

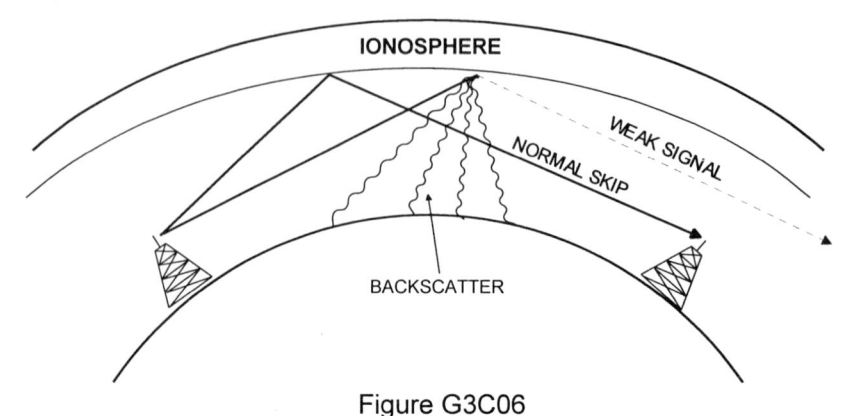

Figure G3C06

In the HF bands, below 10-meters, a one-hop sky wave returning from the ionosphere strikes the earth's surface and, depending on the type of surface, may be scattered in a direction back toward the transmitter. It thus fills in the skip zone with a weak signal. This type of scatter is called BACKSCATTER. It is not a dependable propagation mode, but occurs often enough to be used. Backscatter signals are generally weak and distorted and have a watery or echo type of sound.

G3C07 What makes HF scatter signals often sound distorted?
A. Auroral activity and changes in the earth's magnetic field
B. Propagation through ground waves that absorb much of the signal
C. The state of the E-region at the point of refraction
D. Energy scattered into the skip zone through several radio-wave paths

The answer is D. Scatter waves are invariably distorted and weak because they are returned to earth through several paths. This multipath effect makes the signal enter the skip zone via several paths. The signals arriving from several paths will not be in phase, so distortion will occur. See figure G3C06 and question G3CO6.

G3C08 Why are HF scatter signals usually weak?
A. Only a small part of the signal energy is scattered into the skip zone
B. Auroral activity absorbs most of the signal energy
C. Propagation through ground waves absorbs most of the signal energy
D. The F region of the ionosphere absorbs most of the signal energy

The answer is A. See question G3C06 and figure G3C06.

G3C09 What type of radio-wave propagation allows a signal to be detected at a distance too far for ground-wave propagation but too near for normal sky-wave propagation?
A. Ground wave
B. Scatter
C. Sporadic-E skip
D. Short-path skip

The answer is B. More specifically, it would be backscatter. See answer to question G3C06.

G3C10 When does scatter propagation on the HF bands most often occur?
A. When the sunspot cycle is at a minimum and D-region absorption is high
B. At night
C. When the F1 and F2 regions are combined
D. When communicating on frequencies above the maximum usable frequency (MUF)

The answer is D. See answer to question G3C06.

G3C11 What type of signal fading occurs when two or more parts of a

radio wave follow different paths?
A. Multipath interference
B. Multimode interference
C. Selective iInterference
D. Ionospheric interference
   The answer is A. See answer to question G3C07.

SUBELEMENT G4
AMATEUR RADIO PRACTICES
[5 exam questions - 5 groups]

G4A Two-tone test; electronic TR switch, amplifier neutralization

G4A01 What kind of input signal is used to test the amplitude linearity of a single-sideband phone transmitter while viewing the output on an oscilloscope?
A. Normal speech
B. An audio-frequency sine wave
C. Two audio-frequency sine waves
D. An audio-frequency square wave

The answer is C. A two-tone test signal is used. The two tones are two audio frequency signals of equal amplitude and typically 1 kHz apart in frequency. They must also be within the audio frequency range of the transmitter. Each of the tones must be pure tones and not harmonically related. If your transmitter and amplifier are linear, you will have a stable display on the oscilloscope.

G4A02 When testing the amplitude linearity of a single-sideband transmitter, what kind of audio tones are fed into the microphone input and on what kind of instrument is the output observed?
A. Two harmonically related tones are fed in, and the output is observed on an oscilloscope
B. Two harmonically related tones are fed in, and the output is observed on a distortion analyzer
C. Two non-harmonically related tones are fed in, and the output is observed on an oscilloscope
D. Two non-harmonically related tones are fed in, and the output is observed on a distortion analyzer

The answer is C. Two-tone testing is a common method for determining amplifier linearity. The two tones must be sine wave, non-harmonically related tones. They should be of equal amplitude, free of harmonics and approximately 1000 Hz apart. The two tones should be low enough in frequency to fit inside the audio passband of the SSB transmitter.

The two audio tones are mixed together and fed to the transmitter microphone input. The output is displayed on an oscilloscope where the amplitude of the test signal output, along with the distortion products, can be seen and measured.

Figure G4A02 illustrates oscilloscope patterns using a two-tone test signal. A normal pattern is shown at A. B shows a flattened pattern due to overdrive. C shows a pattern resulting from excessive bias.

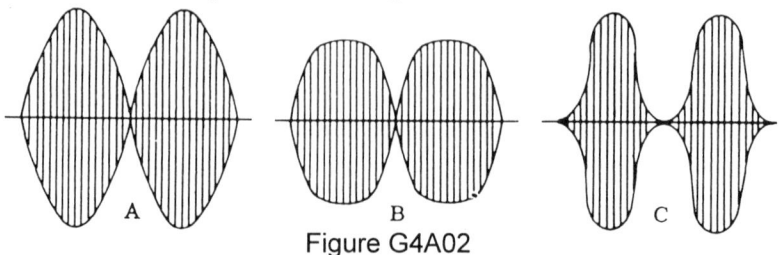

Figure G4A02

G4A03 What audio frequencies are used in a two-tone test of the linearity of a single-sideband phone transmitter?
A. 20 Hz and 20 kHz tones must be used
B. 1200 Hz and 2400 Hz tones must be used
C. Any two audio tones may be used, but they must be within the transmitter audio passband, and must be harmonically related
D. Any two audio tones may be used, but they must be within the transmitter audio passband, and should not be harmonically related
The answer is D. See questions G4A01 and G4A02.

G4A04 What measurement can be made of a single-sideband phone transmitter's amplifier by performing a two-tone test using an oscilloscope?
A. Its percent of frequency modulation
B. Its percent of carrier phase shift
C. Its frequency deviation
D. Its linearity
The answer is D. See questions G4A01 and G4A02. The purpose of the two-tone test is primarily to test and analyze the SSB transmitter for various types of distortion, including the signal-to-distortion ratio and linearity. An oscilloscope is used to observe the signal at the output of the transmitter.

G4A05 At what point in an HF transceiver block diagram would an electronic TR switch normally appear?
A. Between the transmitter and low-pass filter
B. Between the low-pass filter and antenna
C. At the antenna feed point
D. At the power supply feed point
The answer is A. The modern amateur transmitter uses an electronic solid state switch that is very fast, silent and efficient to switch from transmit to receive. One disadvantage of a TR (transmit receive) switch is that it may

generate harmonics that will cause interference. Therefore, it is a good idea to use a low-pass filter between the TR switch and the antenna to eliminate these harmonics. See figure G4A05 for the proper placement of a TR switch in an amateur station.

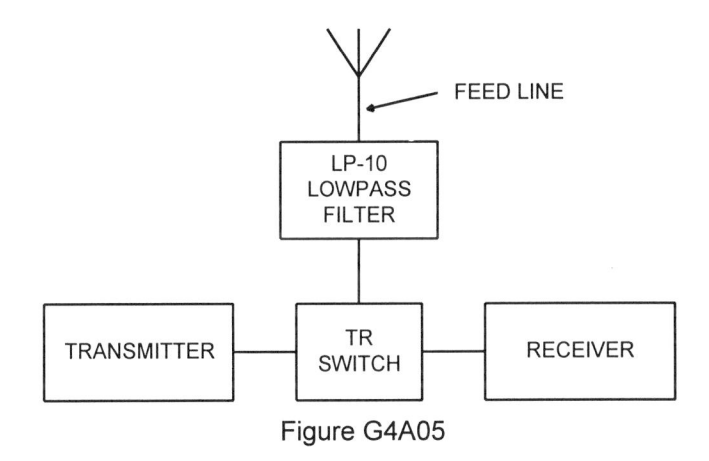

Figure G4A05

G4A06 Why is an electronic TR switch preferable to a mechanical one?
A. It allows greater receiver sensitivity
B. Its circuitry is simpler
C. It has a higher operating speed
D. It allows cleaner output signals

The answer is C. An electronic TR switch will allow much faster operation than a manual switch. It also permits the station to be used with full break-in capability on CW and allows VOX on speech. Electronic switches also eliminate any intermittent problems found with conventional switches and relays.

G4A07 As a power amplifier is tuned, what reading on its grid-current meter indicates the best neutralization?
A. A minimum change in grid current as the output circuit is changed
B. A maximum change in grid current as the output circuit is changed
C. Minimum grid current
D. Maximum grid current

The answer is A. The grid-current meter can be used as a neutralizing indicator because when the amplifier is properly neutralized the change in grid current will be minimum as the output circuit is changed.

G4A08 Why is neutralization necessary for some vacuum-tube amplifiers?
A. To reduce the limits of loaded Q

B. To reduce grid-to-cathode leakage
C. To cancel AC hum from the filament transformer
D. To cancel oscillation caused by the effects of interelectrode capacitance

The answer is D. An RF stage must be neutralized in order to prevent it from oscillating and emitting spurious signals. It oscillates because of the feedback from the plate circuit to the grid circuit, through the plate-grid capacitance of the tube. When RF amplifier stages are designed, a small capacitor is added in such a way that it neutralizes or cancels out the existing interelectrode capacity, thereby preventing the stage from oscillating by eliminating positive feedback.

G4A09 In a properly neutralized RF amplifier, what type of feedback is used?
A. 5%
B. 10%
C. Negative
D. Positive

The answer is C. In an improperly neutralized RF amplifier, positive feedback (some of the output feeds into the input) will cause the amplifier to oscillate. If there is positive feedback from the plate circuit to the grid circuit, through the plate-grid capacitance, the amplifier must be neutralized by adjusting the neutralizing capacitor for minimum change in grid current.

G4A10 What does a neutralizing circuit do in an RF amplifier?
A. It controls differential gain
B. It cancels the effects of positive feedback
C. It eliminates AC hum from the power supply
D. It reduces incidental grid modulation

The answer is B. See questions G4A08 and G4A09.

G4A11 What is the reason for neutralizing the final amplifier stage of a transmitter?
A. To limit the modulation index
B. To eliminate self oscillations
C. To cut off the final amplifier during standby periods
D. To keep the carrier on frequency

The answer is B. See questions G4A08 and G4A09.

G4B Test equipment: oscilloscope; signal tracer; antenna noise bridge; monitoring oscilloscope; field-strength meters

---

G4B01 What item of test equipment contains horizontal- and

vertical-channel amplifiers?
A. An ohmmeter
B. A signal generator
C. An ammeter
D. An oscilloscope

The answer is D. The most useful piece of electronic test gear for the amateur is the oscilloscope. It has a horizontal amplifier that amplifies the signal going to the horizontal plates of the oscilloscope and a vertical amplifier that amplifies the signal going to the vertical plates. The oscilloscope can display on the CRT (cathode ray tube) the wave shape of a signal. It can display AC or DC. The frequency of the signal to be displayed can be quite high depending upon the type of scope.

G4B02 How would a signal tracer normally be used?
A. To identify the source of radio transmissions
B. To make exact drawings of signal waveforms
C. To show standing wave patterns on open-wire feed lines
D. To identify an inoperative stage in a receiver

The answer is D. A signal tracer consists of a detector and an indicating device (speaker, meter, etc.), connected to an input probe. The signal tracer is applied to different parts or stages of a circuit and indicates whether or not a signal is present and its intensity. In this way, it can identify an inoperative stage. The signal being traced is usually supplied by a signal generator or signal injector. The signal tracer is useful for trouble-shooting defective equipment in an amateur station, particularly receivers.

G4B03 Why would you use an antenna noise bridge?
A. To measure the noise figure of an antenna or other electrical circuit
B. To measure the impedance of an antenna or other electrical circuit
C. To cancel electrical noise picked up by an antenna
D. To tune out noise in a receiver

The answer is B. An antenna noise bridge is an instrument that measures the impedance of an antenna. The noise bridge consists of a wide band noise generator and an RF bridge circuit. A zener diode is used as a noise source. The antenna is connected to the instrument so that it becomes one arm of the bridge. A known reactance and resistance in the bridge become another arm of the bridge. When the RF noise generator is turned on, the known reactance and resistance are adjusted for a null on the detector (which is the receiver). By looking at the reading of the known values at null, we can easily calculate the antenna impedance. See answer to question G4B04.

G4B04 How is an antenna noise bridge normally used?

A. It is connected at an antenna's feed point and reads the antenna's noise figure
B. It is connected between a transmitter and an antenna and is tuned for minimum SWR
C. It is connected between a receiver and an unknown impedance and is tuned for minimum noise
D. It is connected between an antenna and ground and is tuned for minimum SWR

The answer is C. One jack on the noise bridge is connected to the antenna. Another jack on the noise bridge is connected to the receiver. The noise source is turned on and the bridge is adjusted for minimum (null) noise. See answer to question G4B03.

G4B05 What is the best instrument to use to check the signal quality of a CW or single-sideband phone transmitter?
A. A monitoring oscilloscope
B. A field-strength meter
C. A sidetone monitor
D. A signal tracer and an audio amplifier

The answer is A. An oscilloscope will allow you to monitor the signal from the transmitter and check for signal quality. By examining the waveform, we can spot distortion and any other problems in the output signal.

G4B06 What signal source is connected to the vertical input of a monitoring oscilloscope when checking the quality of a transmitted signal?
A. The IF output of a monitoring receiver
B. The audio input of the transmitter
C. The RF signals of a nearby receiving antenna
D. The RF output of the transmitter

The answer is D. When we are observing a signal on an oscilloscope, we want to see the amplitude of the signal waveform versus time. To accomplish this, we connect the output of the transmitter to the vertical input of the oscilloscope. The horizontal is connected to the internal sweep of the oscilloscope.

G4B07 What instrument can be used to determine the horizontal radiation pattern of an antenna?
A. A field-strength meter
B. A grid-dip meter
C. An oscilloscope
D. A signal tracer and an audio amplifier

The answer is A. A field-strength meter is a simple but useful piece of test

AMATEUR RADIO PRACTICES   49

equipment that can be used to tune and adjust antennas. A field-strength meter is an RF receiver that displays the relative level of the signal it receives in watts. A very basic field-strength meter consists of a tuned circuit, a silicon diode, and a DC milliammeter. No power supply is needed and it uses only a short antenna (either a telescoping whip or a random length of wire). It can be tuned to a particular wavelength and is used to measure the field strength at varying distances from the antenna. Measurements made in this manner can provide data on the radiation pattern of the antenna.

G4B08 How is a field-strength meter normally used?
A. To determine the standing-wave ratio on a transmission line
B. To check the output modulation of a transmitter
C. To monitor relative RF output
D. To increase average transmitter output

The answer is C. A typical application for a field-strength meter would be to measure the relative output of your antenna. The field-strength meter is moved to several locations away from the transmitter and is used to measure the relative RF level. These measurements can then be used to define the directivity of the antenna. The signal strength (gain) and antenna directivity form the basic radiation pattern of an antenna.

G4B09 What simple instrument may be used to monitor relative RF output during antenna and transmitter adjustments?
A. A field-strength meter
B. An antenna noise bridge
C. A multimeter
D. A metronome

The answer is A. See questions G4B07 and G4B08

G4B10 If the power output of a transmitter is increased by four times, how might a nearby receiver's S-meter reading change?
A. Decrease by approximately one S unit
B. Increase by approximately one S unit
C. Increase by approximately four S units
D. Decrease by approximately four S units

The answer is B. Increasing the power by a factor of four is a 6 dB power increase. This is derived from the dB power formula:

$$dB = 10\log P2/P1 = 10 \log 4 = 10 \times 0.6021 = 6 \text{ dB}$$

(The log table tells us that the log of 4 is 0.6021)
Each unit on an S-meter is equal to approximately 6 dB. Therefore, a power

increase of four times results in an increase in the S-meter reading of one S-unit.

G4B11 By how many times must the power output of a transmitter be increased to raise the S-meter reading on a nearby receiver from S8 to S9?
A. Approximately 2 times
B. Approximately 3 times
C. Approximately 4 times
D. Approximately 5 times
   The answer is C. See question G4B10

G4C Audio rectification in consumer electronics, RF ground

G4C01 What devices would you install to reduce or eliminate audio-frequency interference to home-entertainment systems?
A. Bypass inductors
B. Bypass capacitors
C. Metal-oxide varistors
D. Bypass resistors
   The answer is B. A strong radio frequency signal will be rectified and detected in an early stage of the audio equipment. This interference can be minimized only by modification of the audio equipment. Some of the things that can be done are: shielding and/or by-passing of the power leads, speaker leads, and other interconnecting leads. Also the first audio stage should be by-passed with a capacitor. A 0.001 mfd. capacitor may be used.

G4C02 What should be done if a properly operating amateur station is the cause of interference to a nearby telephone?
A. Make internal adjustments to the telephone equipment
B. Ask the telephone company to install RFI filters
C. Stop transmitting whenever the telephone is in use
D. Ground and shield the local telephone distribution amplifier
   The answer is B. The amateur operator should suggest that his neighbor notify the telephone company. The telephone company can easily cure the problem by by-passing the telephone microphone and lines with RF by-pass capacitors.

G4C03 What sound is heard from a public-address system if audio rectification of a nearby single-sideband phone transmission occurs?
A. A steady hum whenever the transmitter's carrier is on the air
B. On-and-off humming or clicking

AMATEUR RADIO PRACTICES 51

C. Distorted speech from the transmitter's signals
D. Clearly audible speech from the transmitter's signals
   The answer is C. The simple diode detection that occurs in a public address system is not entirely adequate in detecting an SSB signal. The amateur SSB signal will sound distorted in the output of the public address system. If you are using Morse code, there will be on-and-off humming or clicking in the output of the public address system.

G4C04 What sound is heard from a public-address system if audio rectification of a nearby CW transmission occurs?
A. On-and-off humming or clicking
B. Audible, possibly distorted speech
C. Muffled, severely distorted speech
D. A steady whistling
   The answer is A. See question G4C03

G4C05 How can you minimize the possibility of audio rectification of your transmitter's signals?
A. By using a solid-state transmitter
B. By using CW emission only
C. By ensuring that all station equipment is properly grounded
D. By installing bypass capacitors on all power supply rectifiers
   The answer is C. There is little that the amateur radio operator can do at his station to reduce this type of interference, other than cut his power down considerably, and make sure his equipment is properly grounded. The fault lies with the audio equipment. This type of interference can be minimized by modification of the audio equipment. Some of the things that can be done are: shielding and/or by-passing of the power leads, speaker leads and other interconnecting leads with a .001 mfd. capacitor.
   From a practical point of view, it is not a good idea to work on your neighbor's equipment. You should suggest that they contact the manufacturer of the equipment. The manufacturer may provide information or parts so that a local serviceman would be able to modify the equipment to reduce interference.

G4C06 If your third-floor amateur station has a ground wire running 33 feet down to a ground rod, why might you get an RF burn if you touch the front panel of your HF transceiver?
A. Because the ground rod is not making good contact with moist earth
B. Because the transceiver's heat-sensing circuit is not working to start the cooling fan
C. Because of a bad antenna connection, allowing the RF energy to take an easier path out of the transceiver through you

D. Because the ground wire is a resonant length on several HF bands and acts more like an antenna than an RF ground connection

The answer is D. The formula for wavelength in feet is:

$$WL(\text{in feet}) = 984/\text{freq}(\text{in MHz}).$$

Substituting in the formula, you can see that a 33 ft. ground wire would be a good antenna over some of the HF bands. Ground leads must be kept short so that they cannot act as an antenna and cause RF burns by picking up stray RF energy.

G4C07 What is NOT an important reason to have a good station ground?
A. To reduce the cost of operating a station
B. To reduce electrical noise
C. To reduce interference
D. To reduce the possibility of electric shock

The answer is A. Good grounding around your station is a must. It will reduce electrical noise, interference and the possibility of electric shock. Grounding is part of good electrical practice and should be applied to everything in your station, regardless of cost.

G4C08 What is one good way to avoid stray RF energy in your amateur station?
A. Keep the station's ground wire as short as possible
B. Use a beryllium ground wire for best conductivity
C. Drive the ground rod at least 14 feet into the ground
D. Make a couple of loops in the ground wire where it connects to your station

The answer is A. If your ground lead is resonant on any RF bands it will act as an antenna for stray RF signals. See questions G4C06 and G4C07.

G4C09 Which statement about station grounding is NOT true?
A. Braid from RG-213 coaxial cable makes a good conductor to tie station equipment together into a station ground
B. Only transceivers and power amplifiers need to be tied into a station ground
C. According to the National Electrical Code, there should be only one grounding system in a building
D. The minimum length for a good ground rod is 8 feet

The answer is B. Every piece of electronic equipment should be grounded. A heavy braid like the type used on RG-213 makes a great, flexible ground. See answer G4C07.

AMATEUR RADIO PRACTICES   53

G4C10 Which statement about station grounding is true?
A. The chassis of each piece of station equipment should be tied together with high-impedance conductors
B. If the chassis of all station equipment is connected with a good conductor, there is no need to tie them to an earth ground
C. RF hot spots can occur in a station located above the ground floor if the equipment is grounded by a long ground wire
D. A ground loop is an effective way to ground station equipment

The answer is C. Any length of wire that is resonant on one or more of the HF bands will pick up stray RF. This could cause RF hot spots around your station. See answer G4C07.

G4C11 Which of the following is NOT covered in the National Electrical Code?
A. Minimum conductor sizes for different lengths of amateur antennas
B. The size and composition of grounding conductors
C. Electrical safety inside the ham shack
D. The RF exposure limits of the human body

The answer is D. The National Electrical Code is the "bible" of electrical wiring. It covers wire sizes, materials for different applications and electrical safety from the standpoint of proven methods. It has no provisions for RF exposure to the human body. The U.S. Government publishes a guide to RF and Microwave exposure and what is currently thought to be safe levels of exposure. You can request a copy by writing to the Government Printing Office in Pueblo, Colorado.

G4D Speech processors; PEP calculations; wire sizes and fuses

---

G4D01 What is the reason for using a properly adjusted speech processor with a single-sideband phone transmitter?
A. It reduces average transmitter power requirements
B. It reduces unwanted noise pickup from the microphone
C. It improves voice frequency fidelity
D. It improves signal intelligibility at the receiver

The answer is D. A speech processor reduces the ratio between signal peaks and the average power level when used with an SSB transmitter. This increases the average effective radiated power, which results in improved speech intelligibility at the receiver, especially under poor transmission conditions. All modern world-wide sets have built-in speech processors that can be switched in and out to improve the signal.

G4D02 If a single-sideband phone transmitter is 100% modulated, what

will a speech processor do to the transmitter's power?
A. It will increase the output PEP
B. It will add nothing to the output PEP
C. It will decrease the peak power output
D. It will decrease the average power output

The answer is B. Switching on your speech processor will not increase the output power. It increases the average effective radiated power. See answer to question G4D01.

G4D03 How is the output PEP of a transmitter calculated if an oscilloscope is used to measure the transmitter's peak load voltage across a resistive load?
A. PEP = [(Vp)(Vp)] / (RL)
B. PEP = [(0.707 PEV)(0.707 PEV)] / RL
C. PEP = (Vp)(Vp)(RL)
D. PEP = [(1.414 PEV)(1.414 PEV)] / RL

The answer is B. PEP is Peak Envelope Power. It is a very simple matter to measure the Peak Envelope Voltage using an oscilloscope screen. We then use the basic formula:

$$PEP = [0.707PEV \times 0.707PEV] / R$$

to calculate the Peak Envelope Power, where E equals the average voltage. We multiply the PEV by .707 to get the average voltage.

The Peak Envelope Power of a modulated transmitter is an average value over an RF cycle rather than an absolute Peak value.

G4D04 What is the output PEP from a transmitter if an oscilloscope measures 200 volts peak-to-peak across a 50-ohm resistor connected to the transmitter output?
A. 100 watts
B. 200 watts
C. 400 watts
D. 1000 watts

The answer is A. The formula used to solve this problem is:

$$\text{PEP Output Power} = \frac{(\text{Peak voltage} \times .707)^2}{R(\text{load})}$$

This formula is derived from the basic formula, P = E x E / R. The peak voltage is one-half of the peak-to-peak voltage. The peak voltage is therefore 100 volts. We then substitute the known values into the above PEP output power formula and solve the equation.

AMATEUR RADIO PRACTICES 55

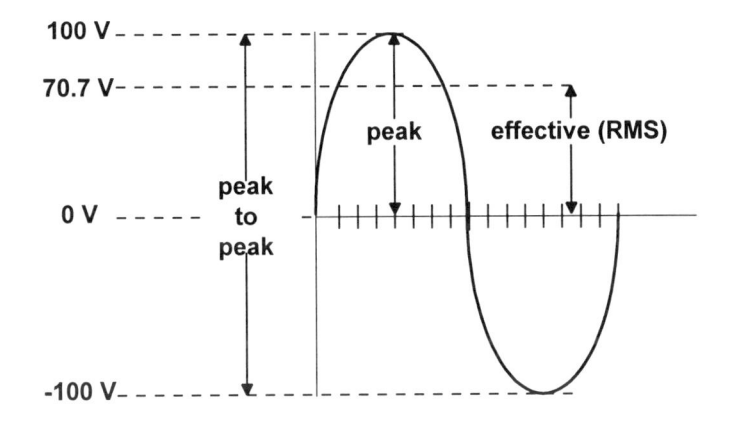

Figure G4D04

$$\text{PEP Output Power} = \frac{(100 \times .707)^2}{50} = \frac{70.7^2}{50} = 100 \text{ watts}$$

G4D05 What is the output PEP from a transmitter if an oscilloscope measures 500 volts peak-to-peak across a 50-ohm resistor connected to the transmitter output?
A. 500 watts
B. 625 watts
C. 1250 watts
D. 2500 watts

The answer is B. See question G4D04.

$$\text{PEP Output Power} = \frac{(250 \times .707)^2}{50} = \frac{176.75^2}{50} = 625 \text{ watts}$$

G4D06 What is the output PEP of an unmodulated carrier transmitter if an average-reading wattmeter connected to the transmitter output indicates 1060 watts?
A. 530 watts
B. 1060 watts
C. 1500 watts
D. 2120 watts

The answer is B. If there is no modulation, the output power would be constant and always at its peak. The PEP output power of an unmodulated carrier transmitter is the same as the average output power.

G4D07 Which wires in a four-conductor line cord should be attached to fuses in a 240-VAC primary (single phase) power supply?
A. Only the "hot" (black and red) wires
B. Only the "neutral" (white) wire
C. Only the ground (bare) wire
D. All wires

The answer is A. House wiring is different from mobile installations, so it is very important that only the hot leads be fused. If the other leads are fused and the fuse opened up, there would be a serious shock hazard.

G4D08 What size wire is normally used on a 15-ampere, 120-VAC household lighting circuit?
A. AWG number 14
B. AWG number 16
C. AWG number 18
D. AWG number 22

The answer is A. This can be arrived at by consulting a table that lists the approved current-carrying capacity of the various sizes and types of wire. If in doubt, always use the next larger size of wire for safety and minimum voltage drop. The smaller AWG numbers indicate a larger diameter wire. The more current that has to be carried, the larger should be the diameter of the wire.

CHART OF WIRE SIZES

Wire Size	Current-Amps (continous duty) single wire	Current-Amps (continous duty) stranded wire
8	70	45
10	55	35
12	40	24
14	32	18
16	22	12
18	15	10

AMATEUR RADIO PRACTICES 57

G4D09 What size wire is normally used on a 20-ampere, 120-VAC household appliance circuit?
A. AWG number 20
B. AWG number 16
C. AWG number 14
D. AWG number 12

The answer is D. See chart used above in question G4D08. Note that a 20 ampere circuit requires a SMALLER NUMBER AWG wire size than the 15 ampere circuit. 15 amps = #14 AWG, 20 amps = #12 AWG.

G4D10 What maximum size fuse or circuit breaker should be used in a household appliance circuit using AWG number 12 wiring?
A. 100 amperes
B. 60 amperes
C. 30 amperes
D. 20 amperes

The answer is D. Since #12 wire should only be used to carry 20 amperes, the fuse should not be any greater than 20 amperes. If a larger fuse is used, it will not limit the current in the line to its proper safe capacity. If we need to increase the current over the 20 amperes, we must increase the wire size to #10 or larger.

G4D11 What maximum size fuse or circuit breaker should be used in a household appliance circuit using AWG number 14 wiring?
A. 15 amperes
B. 20 amperes
C. 30 amperes
D. 60 amperes

The answer is A. #14 wire is used for applications where the current should be limited to 15 amperes so the fuse should be limited to that value of current. See question G4D10.

G4E RF safety

G4E01 Depending on the wavelength of the signal, the energy density of the RF field, and other factors, in what way can RF energy affect body tissue?
A. It heats the tissue
B. It causes radiation poisoning
C. It causes blood flow to stop
D. It produces genetic changes in the tissue

The answer is A. This heating of the tissue can cause severe burns and extensive permanent damage. The severity of the damage will be affected by the RF power level, frequency and length of exposure. The Federal Government publishes exposure limits for RF and Microwave emitting devices but not everything is known about the long term effects to the human body. Good practice would be to keep the exposure to RF and Microwave radiation to a minimum.

G4E02 If you operate your amateur station with indoor antennas, what precautions should you take when you install them?
A. Locate the antennas close to your operating position to minimize feed-line length
B. Position the antennas along the edge of a wall where it meets the floor or ceiling to reduce parasitic radiation
C. Locate the antennas as far away as possible from living spaces that will be occupied while you are operating
D. Position the antennas parallel to electrical power wires to take advantage of parasitic effects

The answer is C. See question G4E01. Keeping the antenna away from living areas will help to limit the exposure to humans.

G4E03 What precaution should you take whenever you make adjustments to the feed system of a parabolic dish antenna?
A. Be sure no one can activate the transmitter
B. Disconnect the antenna-positioning mechanism
C. Point the dish away from the sun so it doesn't concentrate solar energy on you
D. Be sure you and the antenna structure are properly grounded

The answer is A. See question G4E01. Exposure of the body to RF energy is dangerous and you must do everything you can to stop accidental exposure. Safety systems must be used to be sure no one can activate your system while making adjustments to your antenna or feed system.

G4E04 What is one important thing to consider when using an indoor antenna?
A. Use stranded wire to reduce stray RF
B. Ensure that the antenna is as far away from people as possible
C. Use only a Yagi antenna to direct the signals away from people
D. Use as much power as possible to ensure that your signal gets out

The answer is B. See question G4E01. The most important thing to remember is that exposure to RF energy must be limited. With proper placement of your antenna system, any potential danger can be reduced.

AMATEUR RADIO PRACTICES 59

G4E05 Why should a protective fence be placed around the base of a ground-mounted parabolic dish transmitting antenna?
A. To reduce the possibility of persons being harmed by RF energy during transmissions
B. To reduce the possibility that animals will damage the antenna
C. To increase the property value through increased security awareness
D. To protect the antenna from lightning damage and provide a good ground system for the installation

The answer is A. RF safety and the protection of others from harmful RF radiation is the responsibility of the amateur station operator. A ground mounted dish or antenna is where people would most likely be exposed to RF radiation if they were allowed to get too close to the installation. A fence would eliminate the problem.

G4E06 What RF-safety precautions should you take before beginning repairs on an antenna?
A. Be sure you and the antenna structure are grounded
B. Be sure to turn off the transmitter and disconnect the feed line
C. Inform your neighbors so they are aware of your intentions
D. Turn off the main power switch in your house

The answer is B. No precautions are too great to help stop accidental exposure to RF. See the answers to questions G4E04 and G4E05.

G4E07 What precaution should you take when installing a ground--mounted antenna?
A. It should not be installed higher than you can reach
B. It should not be installed in a wet area
C. It should be painted so people or animals do not accidentally run into it
D. It should be installed so no one can come in contact with it

The answer is D. RF safety comes first. See questions G4E05 and G4E06.

G4E08 What precautions should you take before beginning repairs on a microwave feed horn or waveguide?
A. Be sure to wear tight-fitting clothes and gloves to protect your body and hands from sharp edges
B. Be sure the transmitter is turned off and the power source is disconnected
C. Be sure the weather is dry and sunny
D. Be sure propagation conditions are unfavorable for tropospheric ducting

The answer is B. See question G4E06

G4E09 Why should directional high-gain antennas be mounted higher than nearby structures?
A. So they will be dried by the wind after a heavy rain storm
B. So they will not damage nearby structures with RF energy
C. So they will receive more sky waves and fewer ground waves
D. So they will not direct RF energy toward people in nearby structures

The answer is D. RF safety means doing everything possible to limit, reduce or eliminate RF exposure to other people and yourself. Every effort should be used in the antenna system design and layout to keep RF exposure to people at a minimum.

G4E10 For best RF safety, where should the ends and center of a dipole antenna be located?
A. Near or over moist ground so RF energy will be radiated away from the ground
B. As close to the transmitter as possible so RF energy will be concentrated near the transmitter
C. As high as possible to prevent people from coming in contact with the antenna
D. Close to the ground so simple adjustments can be easily made without climbing a ladder

The answer is C. When you transmit, there are high voltages present at the center and ends of the dipole which can cause an RF burn. You must position the antenna so that no one could come in contact with it. RF safety is the responsibility of the amateur station operator.

G4E11 Which property of RF energy is NOT important in estimating the energy's effect on body tissue?
A. The polarization
B. The critical angle
C. The power density
D. The frequency

The answer is B. Polarization, power density and frequency all affect how RF power will affect the body. Critical angle is used in radio wave propagation. See question G3C04.

SUBELEMENT G5
ELECTRICAL PRINCIPLES
[2 exam questions - 2 groups]

G5A Impedance, including matching; resistance, including ohm; reactance, inductance, capacitance and metric divisions of these values

G5A01 What is impedance?
A. The electric charge stored by a capacitor
B. The opposition to the flow of AC in a circuit containing only capacitance
C. The opposition to the flow of AC in a circuit
D. The force of repulsion between one electric field and another with the same charge

The answer is C. The term impedance is the TOTAL opposition to the flow of alternating current. It is used in circuits containing both resistance and reactance. The formula for the impedance of an AC circuit containing both reactance and resistance is:

$$Z = \sqrt{R^2 + X^2}$$

Where Z is Impedance, R is Resistance and X is Reactance.

G5A02 What is reactance?
A. Opposition to DC caused by resistors
B. Opposition to AC caused by inductors and capacitors
C. A property of ideal resistors in AC circuits
D. A large spark produced at switch contacts when an inductor is de-energized

The answer is B. Reactance is the opposition to AC by coils and capacitors. Inductive reactance is the opposition of an inductance to AC. Capacitive reactance is the opposition of a capacitor to AC.

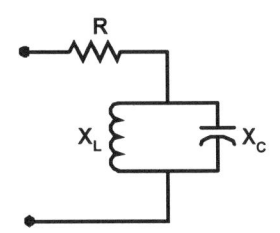

$$X_L = 2 \times \pi \times F \times L$$

$$X_C = \frac{1}{2 \times \pi \times F \times C}$$

Resonant Frequency:

$$F_R = \frac{1}{2 \times \pi \times \sqrt{L \times C}}$$

X_L = Inductive reactance
X_C = Capacitive reactance
$\pi = 3.14$

G5A03 In an inductor, what causes opposition to the flow of AC?
A. Resistance
B. Reluctance
C. Admittance
D. Reactance

The answer is D. An inductor exhibits inductive reactance which will cause opposition to the flow of AC current. The formula for inductive reactance (X_L) is:

$$X_L = 2 \times \pi \times F \times L$$

Where: $\pi = 3.1415$
F = frequency in Hertz
L = inductance in Henries

G5A04 In a capacitor, what causes opposition to the flow of AC?
A. Resistance
B. Reluctance
C. Reactance
D. Admittance

The answer is C. A capacitor exhibits capacitive reactance which will cause opposition to the flow of AC current. The formula for capacitive reactance (X_C) is:

$$X_C = \frac{1}{2 \times \pi \times F \times C}$$

Where: F = frequency in Hertz
C = capacitance in Farads

G5A05 How does a coil react to AC?
A. As the frequency of the applied AC increases, the reactance decreases
B. As the amplitude of the applied AC increases, the reactance increases
C. As the amplitude of the applied AC increases, the reactance decreases
D. As the frequency of the applied AC increases, the reactance increases

The answer is D. A coil is an inductor. It will exhibit inductive reactance in an AC circuit. From the formula in G5A03, the inductive reactance varies directly with the frequency. As the frequency increases, the inductive reactance increases. If the frequency decreases, the inductive reactance also decreases.

G5A06 How does a capacitor react to AC?
A. As the frequency of the applied AC increases, the reactance decreases

B. As the frequency of the applied AC increases, the reactance increases
C. As the amplitude of the applied AC increases, the reactance increases
D. As the amplitude of the applied AC increases, the reactance decreases

The answer is A. A capacitor will exhibit capacitive reactance in an AC circuit. From the formula in G5A04 the capacitive reactance varies inversely with frequency. That is, as the frequency increases, the capacitive reactance decreases; as the frequency decreases, the capacitive reactance increases.

G5A07 When will a power source deliver maximum output to the load?
A. When the impedance of the load is equal to the impedance of the source
B. When the load resistance is infinite
C. When the power-supply fuse rating equals the primary winding current
D. When air wound transformers are used instead of iron-core transformers

The answer is A. The source and load must be matched for maximum power transfer. Impedance matching will allow the maximum energy to be transferred between a source and load.

G5A08 What happens when the impedance of an electrical load is equal to the internal impedance of the power source?
A. The source delivers minimum power to the load
B. The electrical load is shorted
C. No current can flow through the circuit
D. The source delivers maximum power to the load

The answer is D. When source and load are equal, you will have maximum power transfer. See question G5A07.

G5A09 Why is impedance matching important?
A. So the source can deliver maximum power to the load
B. So the load will draw minimum power from the source
C. To ensure that there is less resistance than reactance in the circuit
D. To ensure that the resistance and reactance in the circuit are equal

The answer is A. If you want to get the maximum output from your transmitter to your antenna, the impedances must be matched

G5A10 What unit is used to measure reactance?
A. Mho
B. Ohm
C. Ampere
D. Siemens

The answer is B. The Ohm is the unit of all forms of opposition in a circuit. Resistance, capacitive reactance and inductive reactance are all measured in Ohms.

G5A11 What unit is used to measure impedance?
A. Volt
B. Ohm
C. Ampere
D. Watt

The answer is B. The impedance of a circuit may include resistance, inductive reactance and capacitive reactance all measured in Ohms. See question G5A10.

G5B Decibel, Ohm's law, current and voltage dividers, electrical power calculations and series and parallel components, transformers (either voltage or impedance), sine wave root-mean-square (RMS) value

---

G5B01 A two-times increase in power results in a change of how many dB?
A. 1 dB higher
B. 3 dB higher
C. 6 dB higher
D. 12 dB higher

The answer is B. We can use the power formula to determine this answer.

$$dB = 10 \log \frac{P2}{P1}$$

where: P2 is the higher power
P1 is the smaller power.

We can then substitute "2" for P2 and "1" for P1.

$$dB = 10 \log \frac{2}{1} = 10 \log 2$$

The log table tells us that the log of two is .0301. Therefore:

$$dB = 10 \times .301 = 3 \, dB$$

Thus we see that multiplying the power by 2 gives a 3 dB increase.

G5B02 How can you decrease your transmitter's power by 3 dB?
A. Divide the original power by 1.5
B. Divide the original power by 2

C. Divide the original power by 3
D. Divide the original power by 4

The answer is B. This question is similar to question G5B01. If we multiply the power by 2, we have an increase of 3 dB. Similarly, if we divide the power by 2, we have a decrease of 3 dB. See formula used in question G5B01.

G5B03 How can you increase your transmitter's power by 6 dB?
A. Multiply the original power by 1.5
B. Multiply the original power by 2
C. Multiply the original power by 3
D. Multiply the original power by 4

The answer is D. We can use the formula from question G5B01 and substitute.

$$6 = 10 \log X$$
$$0.6 = \log X$$
$$0.6 = \log 4$$

The log table tells us that the log of 4 is equal to 0.6. Thus, X is equal to 4. Therefore, an increase of 6 dB results from a power increase of 4.

G5B04 If a signal-strength report is "10 dB over S9", what should the report be if the transmitter power is reduced from 1500 watts to 150 watts?
A. S5
B. S7
C. S9
D. S9 plus 5 dB

The answer is C. If we substitute the powers in the dB formula used for question G5B01.

$$dB = 10 \log \frac{1500}{150} = 10 \log 10 = 10 \times 1 = 10 \, dB$$

The log of 10 is 1. Therefore, dB = 10 x 1 = 10 dB
The reduction in signal strength would be 10 dB. We therefore remove the 10 dB from "10 dB over S9" and we are left with the "S9".

G5B05 If a signal-strength report is "20 dB over S9", what should the report be if the transmitter power is reduced from 1500 watts to 15 watts?
A. S5
B. S7
C. S9

D. S9 plus 10 dB

The answer is C. We substitute the powers in the dB formula used for question G5B01.

$$dB = 10 \log \frac{1500}{15} = 10 \log 100 = 10 \times 2 = 20 \text{ dB}$$

The log of 100 is 2. Therefore, dB = 10 x 2 = 20 dB
The reduction in signal strength would be 20 dB. We therefore remove the 20 dB from "20 dB over S9" and we are left with the "S9".

G5B06 If a 1.0-ampere current source is connected to two parallel-connected 10-ohm resistors, how much current passes through each resistor?
A. 10 amperes
B. 2 amperes
C. 1 ampere
D. 0.5 ampere

The answer is D. In a parallel circuit, the sum of the individual branch currents is equal to the source current. Since the two resistors are equal in value, the current divides equally between them and one-half ampere flows through each resistor. This is shown in figure G5B06 below.

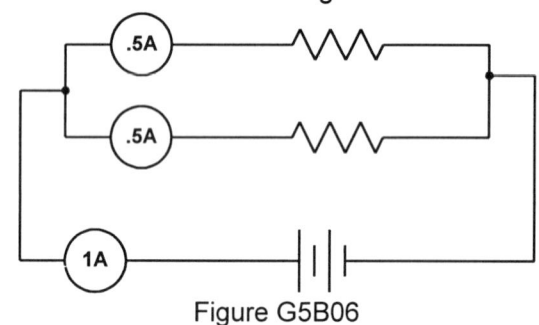

Figure G5B06

G5B07 In a parallel circuit with a voltage source and several branch resistors, how is the total current related to the current in the branch resistors?
A. It equals the average of the branch current through each resistor
B. It equals the sum of the branch current through each resistor
C. It decreases as more parallel resistors are added to the circuit
D. It is the sum of each resistor's voltage drop multiplied by the total number of resistors

The answer is B. In a parallel circuit, the total current is the sum of the currents in each branch. See question G5B06.

ELECTRICAL PRINCIPLES

G5B08 How many watts of electrical power are used if 400 VDC is supplied to an 800-ohm load?
A. 0.5 watts
B. 200 watts
C. 400 watts
D. 320,000 watts

The answer is B. The three formulas that are used to solve problems involving power, voltage and resistance are as follows:

$$P = \frac{E^2}{R} \qquad R = \frac{E^2}{P} \qquad E = \sqrt{P \times R}$$

where: P = watts
E = volts
R = ohms

We use the first formula to solve this problem.

$$P = \frac{E^2}{R} = \frac{400 \times 400}{800} = \frac{160,000}{800} = 200 \text{ watts}$$

G5B09 How many watts of electrical power are used by a 12-VDC light bulb that draws 0.2 amperes?
A. 60 watts
B. 24 watts
C. 6 watts
D. 2.4 watts

The answer is D. The three formulas that are used to solve problems involving power, voltage and current are:

$$P = E \times I \qquad E = \frac{P}{I} \qquad I = \frac{P}{E}$$

We use the first formula to solve this problem.
P(watts) = E(volts) x I(amperes) = 12 x 0.2 = 2.4 watts.

G5B10 How many watts are being dissipated when 7.0 milliamperes flow through 1.25 kilohms?
A. Approximately 61 milliwatts
B. Approximately 39 milliwatts
C. Approximately 11 milliwatts
D. Approximately 9 milliwatts

GENERAL CLASS TEST MANUAL

The answer is A. The three formulas used to solve problems involving power, current and resistance are:

$$P = I^2 \times R \qquad R = \frac{P}{I^2} \qquad I = \sqrt{\frac{P}{R}}$$

Where: P is in watts, I is current in amperes, R is the resistance in ohms. In this problem, we use the first formula. However, we must change milliamperes to amperes and kilohms to ohms before we substitute in the formula.

7.0 mA. = 0.007 A. and 1.25 kilohms = 1250 ohms

$P = I^2 \times R$
$P = 0.007 \times 0.007 \times 1250$
$P = 0.061$ watts or 61 milliwatts

G5B11 What is the voltage across a 500-turn secondary winding in a transformer if the 2250-turn primary is connected to 120 VAC?
A. 2370 volts
B. 540 volts
C. 26.7 volts
D. 5.9 volts

The answer is C. The voltage ratio of a transformer is equal to turns ratio. This is stated mathematically as follows.

$$\frac{T_p}{T_s} = \frac{E_p}{E_s}$$

Where:
T_p is the turns in the primary winding
T_s is the turns in the secondary winding
E_p is the primary voltage
E_s is the secondary voltage

In order to find Es, we must mathematically change the formula so that we can easily work the problem.

$$E_s = \frac{E_p \times T_s}{T_p} = \frac{117 \times 500}{2250} = 26 \text{ volts}$$

G5B12 What is the turns ratio of a transformer to match an audio amplifier having a 600-ohm output impedance to a speaker having a 4-ohm impedance?
A. 12.2 to 1
B. 24.4 to 1
C. 150 to 1
D. 300 to 1

ELECTRICAL PRINCIPLES 69

The answer is A. The turns ratio of a transformer that is used to match the impedance of an amplifier's output to a speaker is found by using the following formula:

$$\text{Turns ratio} = \sqrt{\frac{Z_p}{Z_s}}$$

Where: Zp is the output impedance
Zs is the speaker impedance

We substitute the values given in the problem to arrive at the answer.

$$\text{Turns ratio} = \sqrt{\frac{600}{4}} = 12.25$$

G5B13 What is the impedance of a speaker that requires a transformer with a turns ratio of 24 to 1 to match an audio amplifier having an output impedance of 2000 ohms?
A. 576 ohms
B. 83.3 ohms
C. 7.0 ohms
D. 3.5 ohms

The answer is D. The basic formula showing the relationship between the turns ratio of an impedance matching transformer and the impedances to be matched, is:

$$TR = \sqrt{\frac{Z_p}{Z_s}}$$

Where: TR is the turns ratio
Zp is the output impedance
Zs is the speaker impedance

Since we are interested in finding Zs, we must solve for Zs. Then substitute the values given in the problem.

$$TR^2 = \frac{Z_p}{Z_s} \quad Z_s = \frac{Z_p}{TR^2} = \frac{2000}{24^2} = \frac{2000}{576} = 3.47 \text{ ohms}$$

The impedance of the speaker (Zs) is 3.47 ohms.

G5B14 A DC voltage equal to what value of an applied sine-wave AC voltage would produce the same amount of heat over time in a resistive element?
A. The peak-to-peak value
B. The RMS value

C. The average value
D. The peak value

The answer is B. The DC voltage that will produce the same amount of heat, over time, in a resistive element, as an applied sine-wave AC voltage, is known as the ROOT MEAN SQUARE (RMS) voltage. It is also referred to as the EFFECTIVE value of the voltage. The effective value is equal to .707 multiplied by the peak value of the sine-wave voltage, commonly called heating value.

G5B15 What is the peak-to-peak voltage of a sine wave that has an RMS voltage of 120 volts?
A. 84.8 volts
B. 169.7 volts
C. 204.8 volts
D. 339.4 volts

The answer is D. The peak value of a sine wave is equal to 1.414 multiplied by the RMS VOLTAGE. However, the peak value is the value from zero to peak. This is shown below in Figure G5B15. The peak-to-peak value is the value from the very top of the sine wave peak to the very bottom of the sine wave peak. This is equal to 2.828 multiplied by the RMS voltage. The peak-to-peak voltage of this problem is, therefore, equal to: 2.828 x 120 = 339.4 volts.

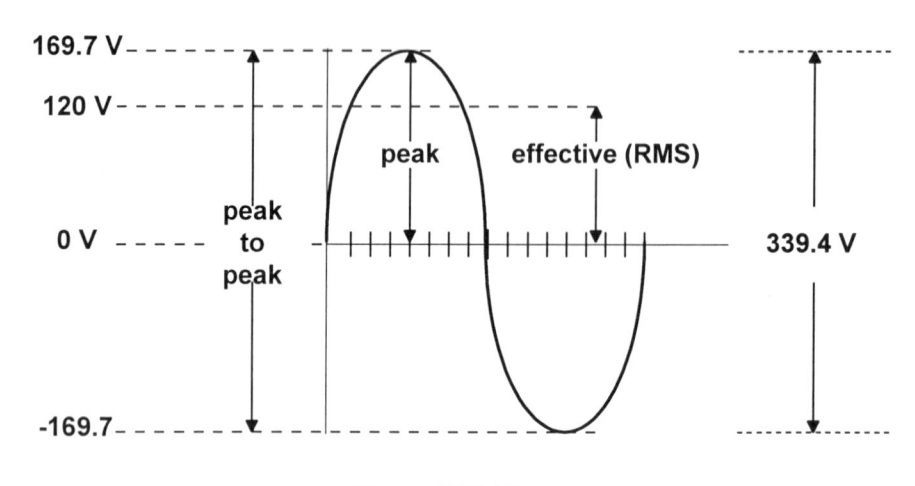

Figure G5B15

G5B16 A sine wave of 17 volts peak is equivalent to how many volts RMS?
A. 8.5 volts
B. 12 volts

C. 24 volts
D. 34 volts

The answer is B. The RMS voltage is equal to .707 multiplied by the peak voltage. We substitute the values given in the problem to find the RMS voltage.

RMS voltage = 0.707 x 17 = 12.02 volts

SUBELEMENT G6
CIRCUIT COMPONENTS
[1 exam question- 1 group]

G6A Resistors, capacitors, inductors, rectifiers and transistors, etc.

G6A01 If a carbon resistor's temperature is increased, what will happen to the resistance?
A. It will increase by 20% for every 10 degrees centigrade
B. It will stay the same
C. It will change depending on the resistor's temperature coefficient rating
D. It will become time dependent

The answer is C. The resistance of carbon resistors decreases with a temperature increase. However, carbon resistors are made with a mixture of carbon and other materials, and the resistance change will depend upon the temperature coefficient of the actual material of which the resistor is made.

G6A02 What type of capacitor is often used in power-supply circuits to filter the rectified AC?
A. Disc ceramic
B. Vacuum variable
C. Mica
D. Electrolytic

The answer is D. Most power-supply filters use electrolytic capacitors. Electrolytic capacitors are ideally suited for power supplies because of their small size for a given amount of capacity. Power supply filters require capacitors with high capacitance values for proper filtering.

G6A03 What type of capacitor is used in power-supply circuits to filter transient voltage spikes across the transformer's secondary winding?
A. High-value
B. Trimmer
C. Vacuum variable
D. Suppressor

The answer is D. Suppressor capacitors are types of capacitors that have high dielectric withstanding voltage ratings so they can resist damage from high voltage spikes. Many years ago, wax paper capacitors were used for their self-healing and high voltage ratings. Today, special surge and spike suppressors are avaiable.

CIRCUIT COMPONENTS

G6A04 Where is the source of energy connected in a transformer?
A. To the secondary winding
B. To the primary winding
C. To the core
D. To the plates

The answer is B. The primary winding of a transformer is connected to the source of AC energy. The secondary winding of the transformer is connected to the load.

G6A05 If no load is attached to the secondary winding of a transformer, what is current in the primary winding called?
A. Magnetizing current
B. Direct current
C. Excitation current
D. Stabilizing current

The answer is A. When there is no load on a transformer you have a basic electro-magnet. Magnetizing current flows in the primary per the following formula.

$$I(amperes) = \frac{E(volts)}{Z(ohms)}$$

Where E is the supply voltage to the primary and Z is the impedance of the primary.

G6A06 What is the peak-inverse-voltage rating of a power-supply rectifier?
A. The maximum transient voltage the rectifier will handle in the conducting direction
B. 1.4 times the AC frequency
C. The maximum voltage the rectifier will handle in the non-conducting direction
D. 2.8 times the AC frequency

The answer is C. The peak-inverse-voltage (PIV) rating is the maximum voltage that can be placed across the diode (rectifier) in its reverse polarity state (anode is negative with respect to cathode). If PIV is exceeded, there may be a breakdown which may destroy the diode.

G6A07 What are the two major ratings that must not be exceeded for silicon-diode rectifiers used in power-supply circuits?
A. Peak inverse voltage; average forward current
B. Average power; average voltage
C. Capacitive reactance; avalanche voltage
D. Peak load impedance; peak voltage

The answer is A. The life of silicon-diode rectifiers depends on the major ratings not being exceeded. PIV (peak inverse voltage) and average forward current (average current drawn by the load) are ratings that should not be exceeded.

G6A08 Why should a resistor and capacitor be wired in parallel with power-supply rectifier diodes?
A. To equalize voltage drops and guard against transient voltage spikes
B. To ensure that the current through each diode is about the same
C. To smooth the output waveform
D. To decrease the output voltage

The answer is A. It is very common in transmitter power supplies to find diodes wired in series in order to obtain a high peak-inverse voltage. When this is done, a resistor is placed across each diode to equalize the voltage drops across the diodes. A capacitor may also be placed across each diode to protect it from high voltage transients and "spikes".

G6A09 What is the output waveform of an unfiltered full-wave rectifier connected to a resistive load?
A. A series of pulses at twice the frequency of the AC input
B. A series of pulses at the same frequency as the AC input
C. A sine wave at half the frequency of the AC input
D. A steady DC voltage

The answer is A. Figure G6A09 below illustrates a full wave rectifier, together with the AC waveform across the transformer secondary, and the output waveform across the load. The output waveform is a pulsating DC wave. Note the entire waveform is above the 0-volt line; therefore, it is a DC voltage.

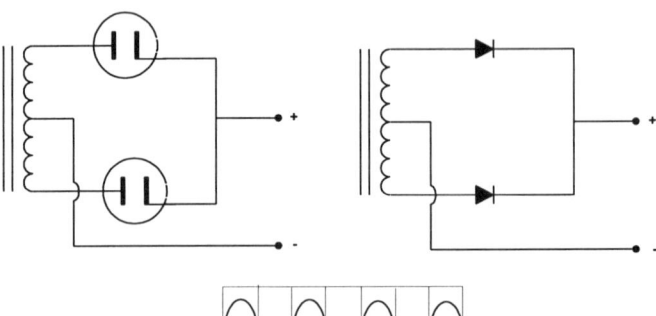

A. Input wave-form
B. Output wave-form

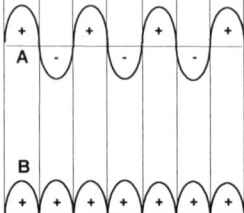

Figure G6A09

CIRCUIT COMPONENTS

G6A10 A half-wave rectifier conducts during how many degrees of each cycle?
A. 90 degrees
B. 180 degrees
C. 270 degrees
D. 360 degrees

The answer is B. Figure G6A10 below illustrates a half-wave rectifier, together with the AC waveform across the transformer secondary, and the output waveform across the load. Note that one-half of the input wave is reproduced. Therefore, half of 360 degrees or 180 degrees is utilized.

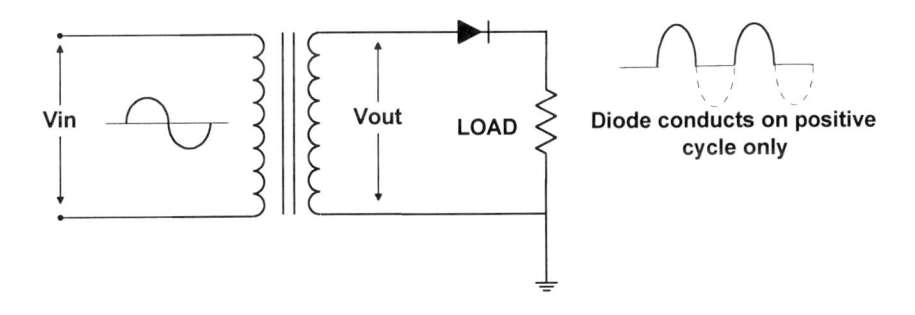

Figure G6A10

G6A11 A full-wave rectifier conducts during how many degrees of each cycle?
A. 90 degrees
B. 180 degrees
C. 270 degrees
D. 360 degrees

The answer is D. We can see in figure G6A09 that the output current flows during the entire cycle. The negative halves of the input waveform are merely inverted. Therefore, 360 degrees of each cycle are utilized in a full-wave rectifier system.

SUBELEMENT G7
PRACTICAL CIRCUITS
[1 exam question - 1 group]

G7A Power supplies and filters; single-sideband transmitters and recievers

G7A01 What safety feature does a power-supply bleeder resistor provide?
A. It improves voltage regulation
B. It discharges the filter capacitors
C. It removes shock hazards from the induction coils
D. It eliminates ground-loop current

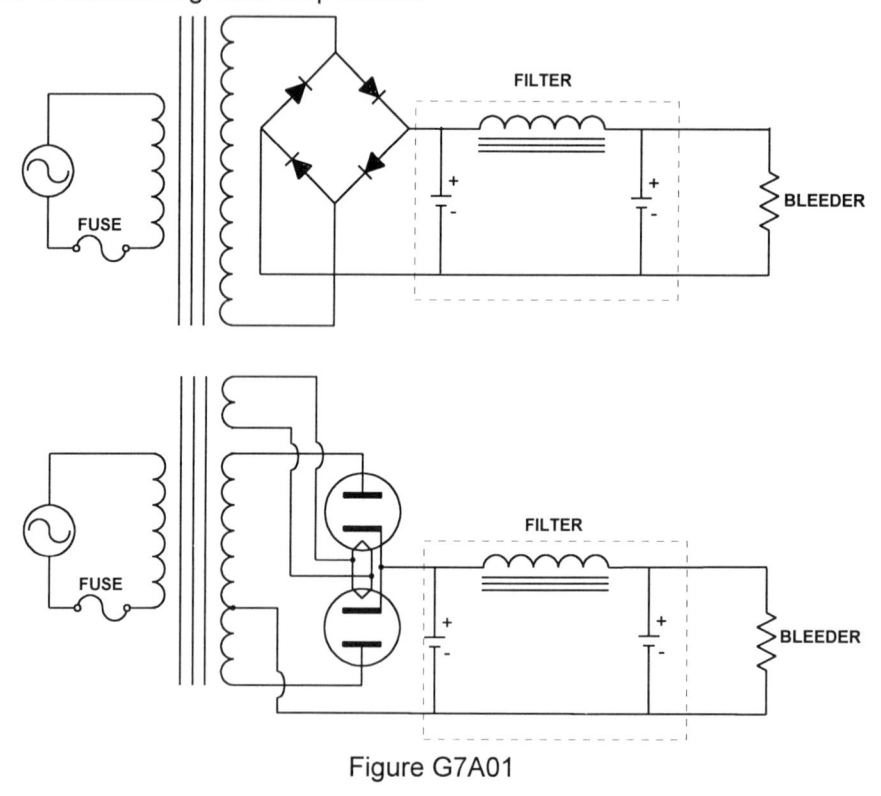

Figure G7A01

The answer is B. When the power supply is operating, the internal filter capacitors will charge up to the power supply voltage. The filter capacitors can retain the charge for quite a while after the supply is turned off creating a safety hazard. A resistor called a bleeder resistor is connected across the output of the power supply allowing the filter capacitors to discharge. See figure G7A01 above.

G7A02 Where is a power-supply bleeder resistor connected?
A. Across the filter capacitor
B. Across the power-supply input
C. Between the transformer primary and secondary windings
D. Across the inductor in the output filter

The answer is A. A bleeder resistor is placed across the output of the power supply. See figure G7A01 above.

G7A03 What components are used in a power-supply filter network?
A. Diodes
B. Transformers and transistors
C. Quartz crystals
D. Capacitors and inductors

The answer is D. A power supply filter network usually consists of one or more capacitors and one or more inductors. Figure G7A01 shows how a practical filter is connected in a power supply.

G7A04 What should be the peak-inverse-voltage rating of the rectifier in a full-wave power supply?
A. One-quarter the normal output voltage of the power supply
B. Half the normal output voltage of the power supply
C. Equal to the normal output voltage of the power supply
D. Double the normal peak output voltage of the power supply

The answer is D. In the case of a full wave rectifier, the peak inverse-voltage is the peak voltage of the entire secondary winding of the transformer. The peak-inverse-voltage is twice the normal peak output voltage of the power supply because the output voltage is derived from one-half of the entire secondary winding.

G7A05 What should be the peak-inverse-voltage rating of the rectifier in a half-wave power supply?
A. One-quarter to one-half the normal peak output voltage of the power supply
B. Half the normal output voltage of the power supply
C. Equal to the normal output voltage of the power supply
D. One to two times the normal peak output voltage of the power supply

The answer is D. See answer G7A04. If the power supply has a resistive or inductive load, the peak-inverse-voltage is equal to the peak voltage of the entire secondary winding of the transformer, which is approximately equal to the normal peak output voltage of the power supply. If the output of the power supply has a capacitive load, the peak-inverse-voltage should be approximately twice the peak output voltage of the power supply or twice the peak voltage of the secondary winding of the transformer. This is because the filter capacitor charges up to the peak value of the secondary winding of the transformer.

G7A06 What should be the impedance of a low-pass filter as compared to the impedance of the transmission line into which it is inserted?
A. Substantially higher
B. About the same
C. Substantially lower
D. Twice the transmission line impedance

The answer is B. For proper operation, the impedance of the lowpass filter should equal the impedance of the transmission line into which it is inserted.

G7A07 In a typical single-sideband phone transmitter, what circuit processes signals from the balanced modulator and sends signals to the mixer?
A. Carrier oscillator
B. Filter
C. IF amplifier
D. RF amplifier

The answer is B. See the block diagram of heterodyne type SSB transmitter, figure G7A07, below.

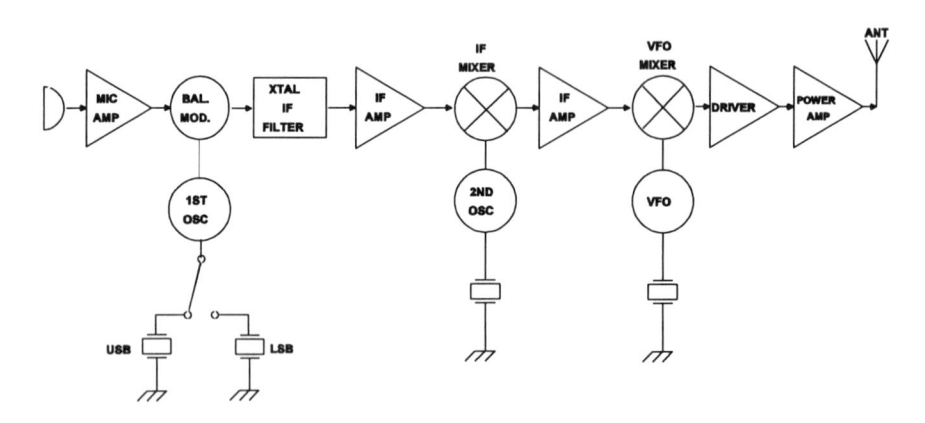

Figure G7A07

G7A08 In a single-sideband phone transmitter, what circuit processes signals from the carrier oscillator and the speech amplifier and sends

PRACTICAL CIRCUITS

signals to the filter?
A. Mixer
B. Detector
C. IF amplifier
D. Balanced modulator

The answer is D. See the block diagram of transmitter, figure G7A07, above.

G7A09 In a single-sideband phone superheterodyne receiver, what circuit processes signals from the RF amplifier and the local oscillator and sends signals to the IF filter?
A. Balanced modulator
B. IF amplifier
C. Mixer
D. Detector

The answer is C. See block diagram of superheterodyne receiver, figure G7A09, below.

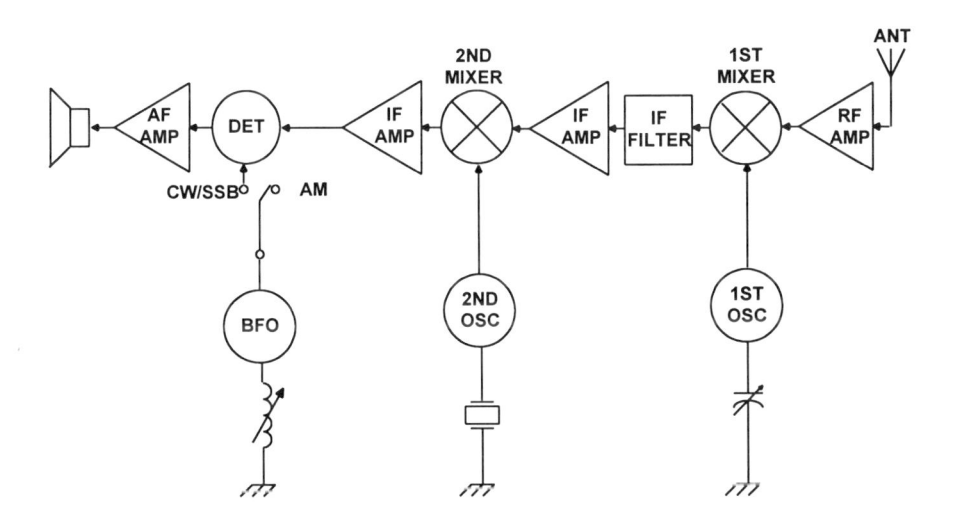

Figure G7A09

G7A10 In a single-sideband phone superheterodyne receiver, what circuit processes signals from the IF amplifier and the BFO and sends signals to the AF amplifier?
A. RF oscillator
B. IF filter
C. Balanced modulator

D. Detector
The answer is D. See the block diagram, figure G7A09.

G7A11 In a single-sideband phone superheterodyne receiver, what circuit processes signals from the IF filter and sends signals to the detector?
A. RF oscillator
B. IF amplifier
C. Mixer
D. BFO
The answer is B. See the block diagram, figure G7A09.

SUBELEMENT G8
SIGNALS AND EMISSIONS
[2 exam questions - 2 groups]

G8A Signal information, AM, FM, single and double sideband and carrier, bandwidth, modulation envelope, deviation, overmodulation

G8A01 What type of modulation system changes the amplitude of an RF wave for the purpose of conveying information?
A. Frequency modulation
B. Phase modulation
C. Amplitude-rectification modulation
D. Amplitude modulation
   The answer is D. In amplitude modulation, the audio signal to be transmitted is superimposed onto the RF carrier by varying the amplitude of the RF carrier in accordance with the audio signal.

G8A02 What type of modulation system changes the phase of an RF wave for the purpose of conveying information?
A. Pulse modulation
B. Phase modulation
C. Phase-rectification modulation
D. Amplitude modulation
   The answer is B. Phase modulation is slightly different from frequency modulation. In frequency modulation we directly alter the frequency of the RF carrier in accordance with the audio. In phase modulation we alter the phase of the carrier which in turn alters the frequency of the carrier. Phase modulation is sometimes called indirect FM. Actually, phase modulation and frequency modulation are both types of FM.

G8A03 What type of modulation system changes the frequency of an RF wave for the purpose of conveying information?
A. Phase-rectification modulation
B. Frequency-rectification modulation
C. Amplitude modulation
D. Frequency modulation
   The answer is D. See questions G8A01 and G8A02.

G8A04 What emission is produced by a reactance modulator connected to an RF power amplifier?
A. Multiplex modulation

B. Phase modulation
C. Amplitude modulation
D. Pulse modulation

The answer is B. When the reactance modulator is connected to an AMPLIFIER, the output will be phase modulated. However, when the reactance modulator is connected to an OSCILLATOR, the output will be frequency modulated. Actually, frequency modulation and phase modulation are similar to each other, since one does not exist without the other.

G8A05 In what emission type does the instantaneous amplitude (envelope) of the RF signal vary in accordance with the modulating audio?
A. Frequency shift keying
B. Pulse modulation
C. Frequency modulation
D. Amplitude modulation

The answer is D. See question G8A01.

G8A06 How much is the carrier suppressed below peak output power in a single-sideband phone transmission?
A. No more than 20 dB
B. No more than 30 dB
C. At least 40 dB
D. At least 60 dB

The answer is C. The carrier should be reduced by at least 40 dB below the peak output power for it to be considered suppressed and the sideband to carry maximum power.

G8A07 What is one advantage of carrier suppression in a double-sideband phone transmission?
A. Only half the bandwidth is required for the same information content
B. Greater modulation percentage is obtainable with lower distortion
C. More power can be put into the sidebands
D. Simpler equipment can be used to receive a double-sideband suppressed-carrier signal

The answer is C. A double-sideband suppressed-carrier signal requires only one-third of the power for similar results, compared to full-carrier AM. This is because two-thirds of the power is in the carrier, which is suppressed, and one-third is in the sidebands.

G8A08 Which popular phone emission uses the narrowest frequency bandwidth?
A. Single-sideband

B. Double-sideband
C. Phase-modulated
D. Frequency-modulated

The answer is A. Single-sideband emissions normally have the narrowest bandwidth. In single-sideband, half of the bandwidth of the double-sideband is suppressed.

G8A09 What happens to the signal of an overmodulated single-sideband or double-sideband phone transmitter?
A. It becomes louder with no other effects
B. It occupies less bandwidth with poor high-frequency response
C. It has higher fidelity and improved signal-to-noise ratio
D. It becomes distorted and occupies more bandwidth

The answer is D. The result of overmodulation is excessive bandwidth and distortion. It will cause interference or "splatter" on nearby frequencies.

G8A10 How should the microphone gain control be adjusted on a single-sideband phone transmitter?
A. For full deflection of the ALC meter on modulation peaks
B. For slight movement of the ALC meter on modulation peaks
C. For 100% frequency deviation on modulation peaks
D. For a dip in plate current

The answer is B. By keeping the gain control adjusted for minimum deflection on the ALC (automatic level control) meter, you will not overmodulate and cause interference on nearby frequencies.

G8A11 What is meant by flattopping in a single-sideband phone transmission?
A. Signal distortion caused by insufficient collector current
B. The transmitter's automatic level control is properly adjusted
C. Signal distortion caused by excessive drive
D. The transmitter's carrier is properly suppressed

The answer is C. Flattopping is the name for a distorted output wave form resulting from excessive drive. See patterns below. A: normal pattern B: Pattern showing flattopping.

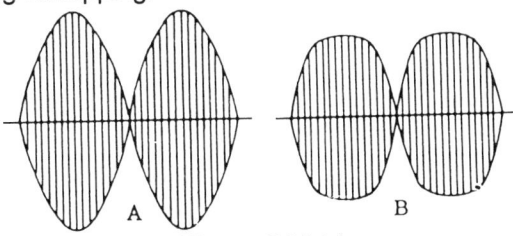

Figure G8A11

G8B Frequency mixing, multiplication, bandwidths, HF data communications

G8B01 What receiver stage combines a 14.25-MHz input signal with a 13.795-MHz oscillator signal to produce a 455-kHz intermediate frequency (IF) signal?
A. Mixer
B. BFO
C. VFO
D. Multiplier
 The answer is A. A mixer stage is used in a receiver to combine frequencies. See question G7A09.

G8B02 If a receiver mixes a 13.800-MHz VFO with a 14.255-MHz received signal to produce a 455-kHz intermediate frequency (IF) signal, what type of interference will a 13.345-MHz signal produce in the receiver?
A. Local oscillator
B. Image response
C. Mixer interference
D. Intermediate interference
 The answer is B. A problem inherent in all mixing systems is image distortion. Whenever two signals are mixed, components are produced at the sum and difference of the two signal frequencies. Filters are used to remove these unwanted out-of-band signals.

G8B03 What stage in a transmitter would change a 5.3-MHz input signal to 14.3 MHz?
A. A mixer
B. A beat frequency oscillator
C. A frequency multiplier
D. A linear translator
 The answer is A. The mixer stage would receive the 5.3 MHz signal and a local oscillator signal of 9 MHz or 19.6 MHz to give us the sum or difference of 14.3 MHz. It could not be a frequency multiplier stage because the output would have been an exact multiple of the input, such as 10.6 or 15.9 MHz.

G8B04 What is the name of the stage in a VHF FM transmitter that selects a harmonic of an HF signal to reach the desired operating frequency?
A. Mixer

B. Reactance modulator
C. Preemphasis network
D. Multiplier
The answer is D. See question G8B03.

G8B05 Why isn't frequency modulated (FM) phone used below 29.5 MHz?
A. The transmitter efficiency for this mode is low
B. Harmonics could not be attenuated to practical levels
C. The bandwidth would exceed FCC limits
D. The frequency stability would not be adequate

The answer is C. The total bandwidth of an FM signal is equal to twice the sum of the deviation and the maximum modulating frequency. For a 5-kHz deviation and a 3-kHz modulating frequency the maximum bandwidth is:

$$\text{Bandwidth} = 2 \times (5 + 3) = 16 \text{ kHz}$$

G8B06 What is the total bandwidth of an FM-phone transmission having a 5-kHz deviation and a 3-kHz modulating frequency?
A. 3 kHz
B. 5 kHz
C. 8 kHz
D. 16 kHz

The answer is D. Bandwidth = 2 x (5 + 3) = 16 kHz
See answer G8B05.

G8B07 What is the frequency deviation for a 12.21-MHz reactance-modulated oscillator in a 5-kHz deviation, 146.52-MHz FM-phone transmitter?
A. 41.67 Hz
B. 416.7 Hz
C. 5 kHz
D. 12 kHz

The answer is B. First we divide the output frequency of the FM transmitter by the oscillator frequency to determine the amount of frequency multiplication of the transmitter.

$$146.52 \text{ MHz divided by } 12.21 \text{ MHz} = 12$$

Then we divide the output frequency deviation by 12 to get the deviation at the oscillator output.

$$5 \text{ kHz } (5000 \text{Hz}) \text{ divided by } 12 = 416.66 \text{ Hz}$$

G8B08 How is frequency shift related to keying speed in an FSK signal?
A. The frequency shift in hertz must be at least four times the keying speed in WPM
B. The frequency shift must not exceed 15 Hz per WPM of keying speed
C. Greater keying speeds require greater frequency shifts
D. Greater keying speeds require smaller frequency shifts

The answer is C. The faster we send digital emmissions and code, the greater the bandwidth they will occupy.

G8B09 What do RTTY, Morse code, AMTOR and packet communications have in common?
A. They are multipath communications
B. They are digital communications
C. They are analog communications
D. They are only for emergency communications

The answer is B.

G8B10 What is the duty cycle required of a transmitter when sending Mode B (FEC) AMTOR?
A. 50%
B. 75%
C. 100%
D. 125%

The answer is C. Mode B employs Forward Error Correction. Data is sent in groups of five characters, then repeated once. This results in continuous transmission and a 100% duty cycle.

G8B11 In what segment of the 20-meter band are most AMTOR operations found?
A. At the bottom of the slow-scan TV segment, near 14.230 MHz
B. At the top of the SSB phone segment, near 14.325 MHz
C. In the middle of the CW segment, near 14.100 MHz
D. At the bottom of the RTTY segment, near 14.075 MHz

The answer is D. These frequency allocations come from the Voluntary HF band Plan for US Operators. These frequency ranges are the result of mutual agreements among amateurs.

SUBELEMENT G9
ANTENNAS AND FEED LINES
[4 exam questions - 4 groups]

G9A Yagi antennas - physical dimensions, impedance matching radiation patterns, directivity and major lobes

G9A01 How can the SWR bandwidth of a parasitic beam antenna be increased?
A. Use larger diameter elements
B. Use closer element spacing
C. Use traps on the elements
D. Use tapered-diameter elements

The answer is A. The bandwidth of an antenna depends on the diameter of the antenna elements and the radiation resistance of the antenna. The larger the diameter of the antenna elements, the greater the bandwidth. The bandwidth will also increase if the radiation resistance is increased. We can increase the radiation resistance of the antenna by increasing the spacing between the elements of the antenna.

G9A02 Approximately how long is the driven element of a Yagi antenna for 14.0 MHz?
A. 17 feet
B. 33 feet
C. 35 feet
D. 66 feet

The answer is B. In a three-element Yagi the director element is approximately 5% shorter than the driven element. The reflector element is approximately 5% longer than the driven element. The formula for the approximate length in feet of the reflector element of a Yagi antenna is:

$$\text{Length (feet)} = \frac{\text{wavelength factor}}{\text{Frequency (MHz)}}$$

The wavelength factor for the reflector element = 490
the driven element = 468
the director element = 458

$$\text{Length} = \frac{468}{14} = 33.43 \text{ feet}$$

G9A03 Approximately how long is the director element of a Yagi antenna for 21.1 MHz?
A. 42 feet
B. 21 feet
C. 17 feet
D. 10.5 feet

The answer is B. The formula for the approximate length in feet of the director element comes from question G9A02.

$$\text{Length} = \frac{458}{21.1} = 21.7 \text{ feet}$$

G9A04 Approximately how long is the reflector element of a Yagi antenna for 28.1 MHz?
A. 8.75 feet
B. 16.6 feet
C. 17.5 feet
D. 35 feet

The answer is C. The formula for the approximate length in feet of the director element comes from question G9A02.

$$\text{Length} = \frac{490}{28.1} = 17.44 \text{ feet}$$

G9A05 Which statement about a three-element Yagi antenna is true?
A. The reflector is normally the shortest parasitic element
B. The director is normally the shortest parasitic element
C. The driven element is the longest parasitic element
D. Low feed-point impedance increases bandwidth

The answer is B. See question G9A02

G9A06 What is one effect of increasing the boom length and adding directors to a Yagi antenna?
A. Gain increases
B. SWR increases
C. Weight decreases
D. Windload decreases

The answer is A. Increasing boom length or adding directors to a three-element Yagi directional antenna will increase gain.

ANTENNAS AND FEEDLINES 89

G9A07 What are some advantages of a Yagi with wide element spacing?
A. High gain, lower loss and a low SWR
B. High front-to-back ratio and lower input resistance
C. Shorter boom length, lower weight and wind resistance
D. High gain, less critical tuning and wider bandwidth

The answer is D. Increased boom length will increase the spacing between elements, thus increasing the radiation resistance and improving SWR and bandwidth. Adding elements will also increase gain.

G9A08 Why is a Yagi antenna often used for radio communications on the 20-meter band?
A. It provides excellent omnidirectional coverage in the horizontal plane
B. It is smaller, less expensive and easier to erect than a dipole or vertical antenna
C. It helps reduce interference from other stations off to the side or behind
D. It provides the highest possible angle of radiation for the HF bands

The answer is C. The 14 MHz band is used for long distance communications. For the best results it is important to have high transmitting and receiving gain in the direction of the station being worked. The Yagi antenna type is very directional and has high gain in the foward direction.

G9A09 What does "antenna front-to-back ratio" mean in reference to a Yagi antenna?
A. The number of directors versus the number of reflectors
B. The relative position of the driven element with respect to the reflectors and directors
C. The power radiated in the major radiation lobe compared to the power radiated in exactly the opposite direction
D. The power radiated in the major radiation lobe compared to the power radiated 90 degrees away from that direction

The answer is C. The "front-to-back" ratio of an antenna is the amount of power radiated in the direction of maximum radiation to the amount of power radiated in the exact opposite (180 degrees) direction.

G9A10 What is the "main lobe" of a Yagi antenna radiation pattern?
A. The direction of least radiation from the antenna
B. The point of maximum current in a radiating antenna element
C. The direction of maximum radiated field strength from the antenna
D. The maximum voltage standing wave point on a radiating element

The answer is C. The RF energy radiated from a directional beam antenna like a three element Yagi is greater in one direction than it is in the other directions. To depict this visually, a graph, called the radiation pattern, is used. See figure G9A10 below.

On the diagram, each egg-shaped protrusion is called a lobe. The largest is the MAJOR or MAIN lobe; all others are referred to as MINOR lobes. The antenna is represented as a small dot in the center of the lobes.

The diagram shows the horizontal pattern, a view looking down on the antenna from above the earth. If lines are drawn outward from the antenna to the edges of the lobes, their lengths are proportional to the signal strengths in those directions. In this example, maximum energy in the MAJOR lobe is to the north. The arrow pointing northeast shows considerably less power in that direction.

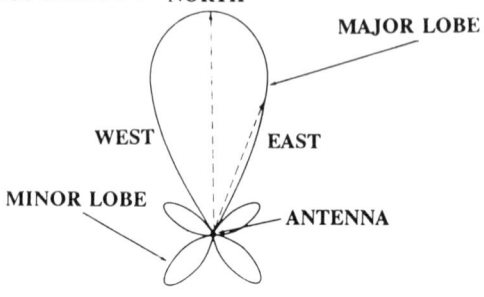

Figure G9A10

G9A11 What is a good way to get maximum performance from a Yagi antenna?
A. Optimize the lengths and spacing of the elements
B. Use RG-58 feed line
C. Use a reactance bridge to measure the antenna performance from each direction around the antenna
D. Avoid using towers higher than 30 feet above the ground

The answer is A. The bandwidth, gain, SWR and transmission resistance all depend on the diameter of the antenna elements, number of elements and the length of the boom. By optimizing all the lengths and spacings of the Yagi antenna, you will improve its performance.

G9B Loop antennas - physical dimensions, impedance matching, radiation patterns, directivity and major lobes

---

G9B01 Approximately how long is each side of a cubical-quad antenna driven element for 21.4 MHz?
A. 1.17 feet
B. 11.7 feet
C. 47 feet
D. 469 feet

ANTENNAS AND FEEDLINES

The answer is B. We use the following formula for determining the total length of the driven element of a cubical-quad antenna:
The wavelength factor for the driven element = 1005
The wavelength factor for the reflector element = 1030

$$\text{Length (feet)} = \frac{\text{wavelength factor}}{\text{Frequency (MHz)}} = \frac{1005}{21.4} = 46.96 \text{ feet}$$

We then divide 46.96 feet by 4 for a quad antenna and we obtain 11.74 feet, which is the length of each side of the driven element of the quad antenna. See appendix for antenna formulas.

G9B02 Approximately how long is each side of a cubical-quad antenna driven element for 14.3 MHz?
A. 17.6 feet
B. 23.4 feet
C. 70.3 feet
D. 175 feet

The answer is A. We use the same formula as in answer G9B01 for determining the total length of the driven element of a cubical-quad antenna:

$$\text{Length (feet)} = \frac{1005}{14.3} = 70.28 \text{ feet}$$

We then divide 70.28 feet by 4 for a quad antenna and we obtain 17.57 feet, which is the length of each side of the driven element of the quad antenna.

G9B03 Approximately how long is each side of a cubical-quad antenna reflector element for 29.6 MHz?
A. 8.23 feet
B. 8.7 feet
C. 9.7 feet
D. 34.8 feet

The answer is B. We use the same formula for determining the total length of the driven element of a cubical-quad antenna but insert the wavelength factor for a reflector element:

$$\text{Length (feet)} = \frac{1030}{29.60} = 34.80 \text{ feet}$$

We then divide 34.80 feet by 4 for a quad antenna and we obtain 8.7 feet, which is the length of each side of the reflector element of the quad antenna.

G9B04 Approximately how long is each leg of a symmetrical delta-loop antenna driven element for 28.7 MHz?
A. 8.75 feet
B. 11.7 feet
C. 23.4 feet
D. 35 feet

The answer is B. The delta loop antenna is similar to the quad antenna in that the total length of the elements of both of them is a full wavelength. However, the delta loop antenna has three sides whereas the quad antenna has four sides. The wavelength factor for the delta loop antenna, driven and reflector elements, is the same as the cubical quad antenna. See formula for question G9B01. The total length of the driven element of a delta loop antenna is:

$$\text{Length (feet)} = \frac{1005}{28.70} = 35.02 \text{ feet}$$

We then divide 35.02 feet by 3 for a delta loop antenna and we obtain 11.67 feet, which is the length of each leg of the driven element of the delta loop antenna.

G9B05 Approximately how long is each leg of a symmetrical delta-loop antenna driven element for 24.9 MHz?
A. 10.99 feet
B. 12.95 feet
C. 13.45 feet
D. 40.36 feet

The answer is C. The total length of the driven element of a delta loop antenna is:

$$\text{Length (feet)} = \frac{1005}{24.90} = 40.36 \text{ feet}$$

We then divide 40.36 feet by 3 for a delta loop antenna and we obtain 13.45 feet, which is the length of each leg of the driven element of the delta loop antenna.

G9B06 Approximately how long is each leg of a symmetrical delta-loop

antenna reflector element for 14.1 MHz?
A. 18.26 feet
B. 23.76 feet
C. 24.35 feet
D. 73.05 feet

The answer is C. The total length of the reflector element of a delta loop antenna is:

$$\text{Length (feet)} = \frac{1030}{14.10} = 73.05 \text{ feet}$$

We then divide 73.05 feet by 3 for a delta loop antenna and we obtain 24.35 feet, which is the length of each leg of the reflector element of the delta loop antenna.

G9B07 Which statement about two-element delta loops and quad antennas is true?
A. They compare favorably with a three-element Yagi
B. They perform poorly above HF
C. They perform very well only at HF
D. They are effective only when constructed using insulated wire

The answer is A. A two-element quad or delta loop antenna compares favorably with a three-element Yagi array in terms of gain (see QST, May 1963 and January 1969, for additional information). The quad and delta loop-antennas perform very well at 50 and 144 MHz. A discussion of radiation patterns and gain, for quads vs. Yagis, is presented in May 1968 QST.

G9B08 Compared to a dipole antenna, what are the directional radiation characteristics of a cubical-quad antenna?
A. The quad has more directivity in the horizontal plane but less directivity in the vertical plane
B. The quad has less directivity in the horizontal plane but more directivity in the vertical plane
C. The quad has more directivity in both horizontal and vertical planes
D. The quad has less directivity in both horizontal and vertical planes

The answer is C. The Quad antenna has more gain and directivity in both tha horizontal and vertical directions than a comparable dipole.

G9B09 Moving the feed point of a multielement quad antenna from a side parallel to the ground to a side perpendicular to the ground will have what effect?
A. It will significantly increase the antenna feed-point impedance

B. It will significantly decrease the antenna feed-point impedance
C. It will change the antenna polarization from vertical to horizontal
D. It will change the antenna polarization from horizontal to vertical

The answer is D. Most antennas are vertically or horizontally polarized. The polarization is determined by the position of the radiating element or wire with respect to the earth. Thus, a radiator that is parallel to the earth radiates horizontally, while an antenna at a right angle to the earth (vertical) radiates a vertical wave. If a wire antenna is used slanted to the earth, it will radiate waves that are between vertical and horizontal.

G9B10 What does the term "antenna front-to-back ratio" mean in reference to a delta-loop antenna?
A. The number of directors versus the number of reflectors
B. The relative position of the driven element with respect to the reflectors and directors
C. The power radiated in the major radiation lobe compared to the power radiated in exactly the opposite direction
D. The power radiated in the major radiation lobe compared to the power radiated 90 degrees away from that direction

The answer is C. The definition of the "front-to-back ratio" is the same for all antennas of the directional type. Yagis, quads, delta loops, quad beams, etc. are all examples of directional antennas. Thus, the power radiated in the major radiation lobe direction compared to the power radiated in exactly the opposite direction would give us the "front-to-back ratio". See drawing of delta loop directional antenna showing forward and reverse transmission patterns.

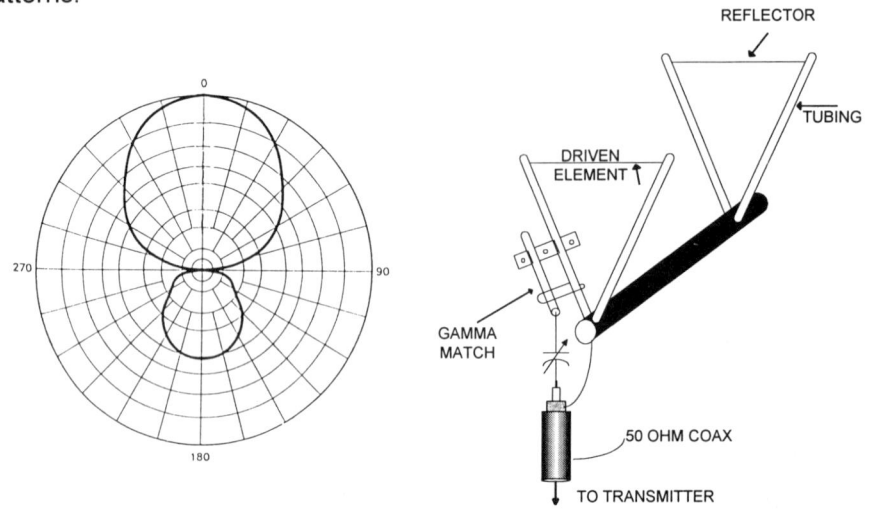

Figure G9B10

ANTENNAS AND FEEDLINES

G9B11 What is the "main lobe" of a delta-loop antenna radiation pattern?
A. The direction of least radiation from an antenna
B. The point of maximum current in a radiating antenna element
C. The direction of maximum radiated field strength from the antenna
D. The maximum voltage standing wave point on a radiating element

The answer is C. The main lobe of all antennas of the directional type, Yagi's, quads, delta loops, quad beams, etc. would be the lobe showing the largest energy emission. RF energy radiated from a beam antenna is greater in one direction than it is in the other direction. To depict this visually, a graph, called the radiation pattern, is used. On the diagram, each egg shaped protrusion is called a lobe. The largest is the major or main lobe; all others are referred to as minor lobes. The antenna is represented as a small dot in the center of the lobe.

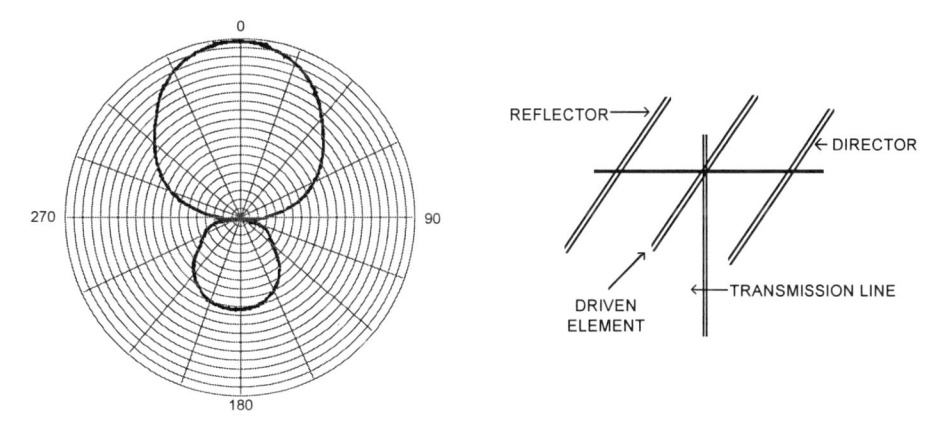

Figure G9B11

G9C Random wire antennas - physical dimensions, impedance matching, radiation patterns, directivity and major lobes; feedpoint impedance of 1/2-wavelength dipole and 1/4-wavelength vertical antennas

---

G9C01 What type of multiband transmitting antenna does NOT require a feed line?
A. A random-wire antenna
B. A triband Yagi antenna
C. A delta-loop antenna
D. A Beverage antenna

The answer is A. Random-length antennas are versatile; they can be used anywhere. But they do have one major disadvantage. Unlike the dipole and vertical antenna, which can be fed directly from the transmitter through coaxial cable, a random-length antenna requires a matching network. The matching network is required because the antenna impedance is not likely to be 50 or 75 ohms.

G9C02 What is one advantage of using a random-wire antenna?
A. It is more efficient than any other kind of antenna
B. It will keep RF energy out of your station
C. It doesn't need an impedance matching network
D. It is a multiband antenna

The answer is D. A properly constructed random-wire antenna can operate with very low loss in the VHF and UHF range.

G9C03 What is one disadvantage of a random-wire antenna?
A. It must be longer than 1 wavelength
B. You may experience RF feedback in your station
C. It usually produces vertically polarized radiation
D. You must use an inverted-T matching network for multiband operation

The answer is B. See answer to question G9C01.

G9C04 What is an advantage of downward sloping radials on a ground-plane antenna?
A. It lowers the radiation angle
B. It brings the feed-point impedance closer to 300 ohms
C. It increases the radiation angle
D. It brings the feed-point impedance closer to 50 ohms

The answer is D. Downward sloping radials increase the feed point impedance closer to 50 ohms. This makes it easier for the antenna to match the impedance of most transmission lines. When the radials are not sloping, the impedance is approximately 30 to 36 ohms.

G9C05 What happens to the feed-point impedance of a ground-plane antenna when its radials are changed from horizontal to downward-sloping?
A. It decreases
B. It increases
C. It stays the same
D. It approaches zero

The answer is B. When the radials are horizontal, the impedance is approximately 30 to 36 ohms. See answer to question G9C04.

G9C06 What is the low-angle radiation pattern of an ideal half-wavelength dipole HF antenna installed parallel to the earth?

ANTENNAS AND FEEDLINES

A. It is a figure-eight at right angles to the antenna
B. It is a figure-eight off both ends of the antenna
C. It is a circle (equal radiation in all directions)
D. It is two smaller lobes on one side of the antenna, and one larger lobe on the other side

The answer is A. The response of a dipole antenna in free space, looking at the plane of the conductor, looks like a figure-eight. See diagram below.

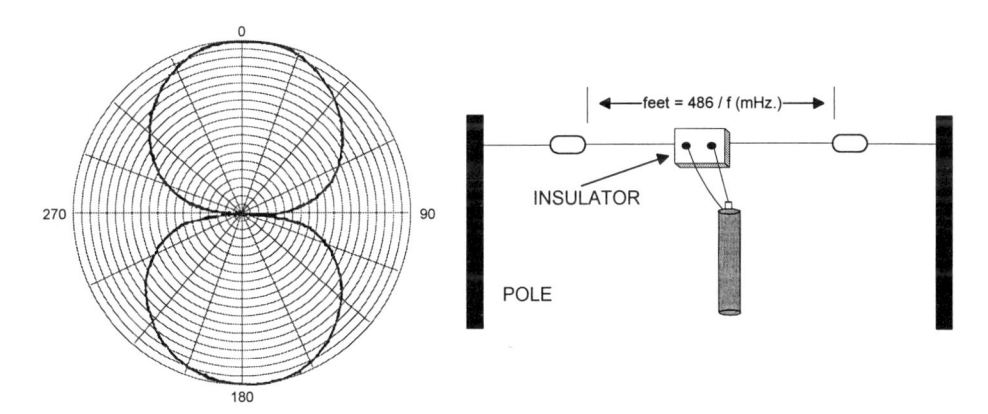

Figure G9C06

G9C07 How does antenna height affect the horizontal (azimuthal) radiation pattern of a horizontal dipole HF antenna?
A. If the antenna is too high, the pattern becomes unpredictable
B. If the antenna is less than one-half wavelength high, reflected radio waves from the ground significantly distort the pattern
C. Antenna height has no effect on the pattern
D. If the antenna is less than one-half wavelength high, radiation off the ends of the wire is eliminated

The answer is B. Many amateurs incorrectly assume that a dipole antenna will exhibit a good pattern at any height above the ground. As the antenna is moved closer to the ground, the radiation pattern deteriorates until it is completly distorted.

G9C08 If a slightly shorter parasitic element is placed 0.1 wavelength away from an HF dipole antenna, what effect will this have on the antenna's radiation pattern?
A. The radiation pattern will not be affected
B. A major lobe will develop in the horizontal plane, parallel to the two elements
C. A major lobe will develop in the vertical plane, away from the ground

D. A major lobe will develop in the horizontal plane, toward the parasitic element

The answer is D. Elements that receive power by induction or radiation from a driven element are called parasitic elements. The parasitic element is called a director when it reinforces the radiation on a line from the driven element and a reflector when the reverse is the case. Parasitic elements are tuned by changing their length.

G9C09 If a slightly longer parasitic element is placed 0.1 wavelength away from an HF dipole antenna, what effect will this have on the antenna's radiation pattern?
A. The radiation pattern will not be affected
B. A major lobe will develop in the horizontal plane, away from the parasitic element, toward the dipole
C. A major lobe will develop in the vertical plane, away from the ground
D. A major lobe will develop in the horizontal plane, parallel to the two elements

The answer is B. See the answer to question C9C08.

G9C10 Where should the radial wires of a ground-mounted vertical antenna system be placed?
A. As high as possible above the ground
B. On the surface or buried a few inches below the ground
C. Parallel to the antenna element
D. At the top of the antenna

The answer is B. Vertical antennas are simple and popular. On the HF bands (80-10 meters) they are commonly used for DX applications. For operation over the 80 to 10-meter bands, the antenna may be at ground level and the radials placed on the ground or just a few inches under. The radial system is the reason these antennas perform so well. The best radial system will use many radials laid out in a circular pattern. Four radials should be considered an absolute minimum.

G9C11 If you are transmitting from a ground-mounted vertical antenna, which of the following is an important reason for people to stay away from it?
A. To avoid skewing the radiation pattern
B. To avoid changes to the antenna feed-point impedance
C. To avoid excessive grid current
D. To avoid exposure to RF radiation

The answer is D. All ground mounted vertical antennas represent an RF radiation hazard and should never be placed where someone could accidentally touch it or be exposed to RF radiation.

ANTENNAS AND FEEDLINES   99

> **G9D** Popular antenna feed lines - characteristic impedance and impedance matching; SWR calculations

G9D01 What factors determine the characteristic impedance of a parallel-conductor antenna feed line?
A. The distance between the centers of the conductors and the radius of the conductors
B. The distance between the centers of the conductors and the length of the line
C. The radius of the conductors and the frequency of the signal
D. The frequency of the signal and the length of the line

The answer is A. The characteristic impedance of an air insulated parallel-conductor transmission line is dependent upon the radius of the conductors and the distance between the conductors. Its formula is:

$Z = 276 \log b / a$ Where: "b" is the distance between conductors and "a" is the radius.

The dielectric used to separate the conductors also affects the impedance.

G9D02 What is the typical characteristic impedance of coaxial cables used for antenna feed lines at amateur stations?
A. 25 and 30 ohms
B. 50 and 75 ohms
C. 80 and 100 ohms
D. 500 and 750 ohms

The answer is B. While most antenna designs will yield a 50 ohm impedance, there are some popular types that yield 75 ohms as well.

G9D03 What is the characteristic impedance of flat-ribbon TV-type twinlead?
A. 50 ohms
B. 75 ohms
C. 100 ohms
D. 300 ohms

The answer is D. The dielectric material and lead spacing used on TV-type twinlead result in 300 ohm impedance over the VHF-UHF range. See the answer to question G9D01.

G9D04 What is the typical cause of power being reflected back down an antenna feed line?
A. Operating an antenna at its resonant frequency

B. Using more transmitter power than the antenna can handle
C. A difference between feed-line impedance and antenna feed-point impedance
D. Feeding the antenna with unbalanced feed line

The answer is C. An antenna transmission-line mismatch increases the SWR of the system. This reduces the efficiency of the antenna system because of the reflected power back to the transmitter.

G9D05 What must be done to prevent standing waves of voltage and current on an antenna feed line?
A. The antenna feed point must be at DC ground potential
B. The feed line must be cut to an odd number of electrical quarter-wavelengths long
C. The feed line must be cut to an even number of physical half-wavelengths long
D. The antenna feed-point impedance must be matched to the characteristic impedance of the feed line

The answer is D. The impedance of the transmission line must be equal to the impedance of the antenna at the point where the feed line connects to the antenna.

G9D06 If a center-fed dipole antenna is fed by parallel-conductor feed line, how would an inductively coupled matching network be used between the two?
A. It would not normally be used with parallel-conductor feed lines
B. It would be used to increase the SWR to an acceptable level
C. It would be used to match the unbalanced transmitter output to the balanced parallel-conductor feed line
D. It would be used at the antenna feed point to tune out the radiation resistance

The answer is C. The inductively coupled matching network is called a balun (BALanced to UNbalanced). It matches the balanced antenna to the unbalanced transmission line. The balun is placed at the point where the transmission line feeds the antenna.

G9D07 If a 160-meter signal and a 2-meter signal pass through the same coaxial cable, how will the attenuation of the two signals compare?
A. It will be greater at 2 meters
B. It will be less at 2 meters
C. It will be the same at both frequencies
D. It will depend on the emission type in use

The answer is A. As the operating frequency increases, the dielectric

ANTENNAS AND FEEDLINES

losses in coaxial feed lines increase. As the operating frequency decreases, the dielectric losses in coaxial feed lines decrease. This is why loss specifications of transmission lines give the frequency at which the losses are measured.

G9D08 In what values are RF feed line losses usually expressed?
A. Bels/1000 ft
B. dB/1000 ft
C. Bels/100 ft
D. dB/100 ft

The answer is D. It is usually expressed in decibels (dB) per unit length, usually dB per 100 feet.

G9D09 What standing-wave-ratio will result from the connection of a 50-ohm feed line to a resonant antenna having a 200-ohm feed-point impedance?
A. 4:1
B. 1:4
C. 2:1
D. 1:2

The answer is A. Standing waves are the graphs of voltage and current that occur along a resonant antenna. Standing waves occur because an RF signal, traveling along an antenna, strikes the end and is reflected back. In traveling back, it either reinforces or cancels the original signal at different points. The maximums and minimums of the voltages and currents appear in the same positions along the wires, hence the name "standing waves". The standing wave ratio is equal to the larger impedance, divided by the smaller impedance:

$$SWR = \frac{200}{50} = \frac{4}{1} \text{ or } 4:1$$

G9D10 What standing-wave-ratio will result from the connection of a 50-ohm feed line to a resonant antenna having a 10-ohm feed-point impedance?
A. 2:1
B. 50:1
C. 1:5
D. 5:1

The answer is D. The standing wave ratio is equal to the larger impedance divided by the smaller impedance. See the answer to question G9D09.

G9D11 What standing-wave-ratio will result from the connection of a 50-ohm feed line to a resonant antenna having a 50-ohm feed-point impedance?
A. 2:1
B. 50:50
C. 0:0
D. 1:1

The answer is D. The standing wave ratio is equal to the larger impedance, divided by the smaller impedance. The SWR is 1:1 and is achieved when the characteristic impedance of the line is equal to the resistance at the antenna feedpoint. This is an ideal situation. You may not be able to meet a 1:1 SWR, but with a little work on your antenna system, you can reach between 1.5:1 to 2:1 SWR.

SAMPLE GENERAL CLASS EXAMINATION

The following sample examination provides an added means of preparation for the actual test and can serve as a gauge of preparedness. The answers are on the page following this exam. Please note that each question is numbered as it appears in the Subelements, making it easier to look up the question and related discussion.

G1A03 What are the frequency limits for General class operators in the 40-meter band (ITU Region 2)? [97.301d]
A. 7025 - 7175 kHz and 7200 - 7300 kHz
B. 7025 - 7175 kHz and 7225 - 7300 kHz
C. 7025 - 7150 kHz and 7200 - 7300 kHz
D. 7025 - 7150 kHz and 7225 - 7300 kHz

Answer: _____

G1B04 Which of the following does NOT need to be true if an amateur station gathers news information for broadcast purposes? [97.113c]
A. The information is more quickly transmitted by amateur radio
B. The information must involve the immediate safety of life of individuals or the immediate protection of property
C. The information must be directly related to the event
D. The information cannot be transmitted by other means

Answer: _____

G1C08 Which of the following standards must be met if FCC type acceptance of an external RF amplifier is required? [97.317c6i]
A. The amplifier must not be able to amplify a 28-MHz signal to more than ten times the input power
B. The amplifier must not be capable of reaching its designed output power when driven with less than 50 watts
C. The amplifier must not be able to be operated for more than ten minutes without a time delay circuit
D. The amplifier must not be able to be modified by an amateur operator

Answer: _____

G1D10 If you are a Technician licensee with a Certificate of Successful Completion of Examination (CSCE) for General privileges, how do you identify your station when transmitting phone emissions on 14.325 MHz? [97.119e2]
A. No special form of identification is needed
B. You may not operate on 14.325 MHz until your new license arrives

C. You must give your call sign, followed by any suitable word that denotes the slant mark and the identifier "AG"
D. You must give your call sign and the location of the VE examination where you obtained the CSCE

Answer: _____

G2A07 What is the most common frequency shift for RTTY emissions in the amateur HF bands?
A. 85 Hz
B. 170 Hz
C. 425 Hz
D. 850 Hz

Answer: _____

G2B02 If a net is about to begin on a frequency which you and another station are using, what should you do?
A. As a courtesy to the net, move to a different frequency
B. Increase your power output to ensure that all net participants can hear you
C. Transmit as long as possible on the frequency so that no other stations may use it
D. Turn off your radio

Answer: _____

G2C11 What are the objectives of the Amateur Auxiliary to the FCC's Field Operations Bureau?
A. To conduct efficient and orderly amateur licensing examinations
B. To encourage amateur self-regulation and compliance with the rules
C. To coordinate repeaters for efficient and orderly spectrum usage
D. To provide emergency and public safety communications

Answer: _____

G3A02 What effect does a sudden ionospheric disturbance have on the daylight ionospheric propagation of HF radio waves?
A. It disrupts higher-latitude paths more than lower-latitude paths
B. It disrupts signals on lower frequencies more than those on higher frequencies
C. It disrupts communications via satellite more than direct communications
D. None, only areas on the night side of the earth are affected

Answer: _____

G3B07 During periods of low solar activity, which frequencies are the least reliable for long-distance communication?
A. Frequencies below 3.5 MHz
B. Frequencies near 3.5 MHz
C. Frequencies on or above 10 MHz
D. Frequencies above 20 MHz

Answer: _____

G3C07 What makes HF scatter signals often sound distorted?
A. Auroral activity and changes in the earth's magnetic field
B. Propagation through ground waves that absorb much of the signal
C. The state of the E-region at the point of refraction
D. Energy scattered into the skip zone through several radio-wave paths

Answer: _____

G4A09 In a properly neutralized RF amplifier, what type of feedback is used?
A. 5%
B. 10%
C. Negative
D. Positive

Answer: _____

G4B02 How would a signal tracer normally be used?
A. To identify the source of radio transmissions
B. To make exact drawings of signal waveforms
C. To show standing wave patterns on open-wire feed lines
D. To identify an inoperative stage in a receiver

Answer: _____

G4C04 What sound is heard from a public-address system if audio rectification of a nearby CW transmission occurs?
A. On-and-off humming or clicking
B. Audible, possibly distorted speech
C. Muffled, severely distorted speech
D. A steady whistling

Answer: _____

G4D09 What size wire is normally used on a 20-ampere, 120-VAC household appliance circuit?

A. AWG number 20
B. AWG number 16
C. AWG number 14
D. AWG number 12

Answer: _____

G4E03 What precaution should you take whenever you make adjustments to the feed system of a parabolic dish antenna?
A. Be sure no one can activate the transmitter
B. Disconnect the antenna-positioning mechanism
C. Point the dish away from the sun so it doesn't concentrate solar energy on you
D. Be sure you and the antenna structure are properly grounded

Answer: _____

G5A05 How does a coil react to AC?
A. As the frequency of the applied AC increases, the reactance decreases
B. As the amplitude of the applied AC increases, the reactance increases
C. As the amplitude of the applied AC increases, the reactance decreases
D. As the frequency of the applied AC increases, the reactance increases

Answer: _____

G5B16 A sine wave of 17 volts peak is equivalent to how many volts RMS?
A. 8.5 volts
B. 12 volts
C. 24 volts
D. 34 volts

Answer: _____

G6A06 What is the peak-inverse-voltage rating of a power-supply rectifier?
A. The maximum transient voltage the rectifier will handle in the conducting direction
B. 1.4 times the AC frequency
C. The maximum voltage the rectifier will handle in the non-conducting direction
D. 2.8 times the AC frequency

Answer: _____

G7A05 What should be the peak-inverse-voltage rating of the rectifier in a half-wave power supply?
A. One-quarter to one-half the normal peak output voltage of the power supply
B. Half the normal output voltage of the power supply
C. Equal to the normal output voltage of the power supply
D. One to two times the normal peak output voltage of the power supply

Answer: _____

G8A10 How should the microphone gain control be adjusted on a single-sideband phone transmitter?
A. For full deflection of the ALC meter on modulation peaks
B. For slight movement of the ALC meter on modulation peaks
C. For 100% frequency deviation on modulation peaks
D. For a dip in plate current

Answer: _____

G8B02 If a receiver mixes a 13.800-MHz VFO with a 14.255-MHz received signal to produce a 455-kHz intermediate frequency (IF) signal, what type of interference will a 13.345-MHz signal produce in the receiver?
A. Local oscillator
B. Image response
C. Mixer interference
D. Intermediate interference

Answer: _____

G9A04 Approximately how long is the reflector element of a Yagi antenna for 28.1 MHz?
A. 8.75 feet
B. 16.6 feet
C. 17.5 feet
D. 35 feet

Answer: _____

G9B07 Which statement about two-element delta loops and quad antennas is true?
A. They compare favorably with a three-element Yagi
B. They perform poorly above HF
C. They perform very well only at HF
D. They are effective only when constructed using insulated wire

Answer: _____

G9C04 What is an advantage of downward sloping radials on a ground-plane antenna?
A. It lowers the radiation angle
B. It brings the feed-point impedance closer to 300 ohms
C. It increases the radiation angle
D. It brings the feed-point impedance closer to 50 ohms

Answer: _____

G9D09 What standing-wave-ratio will result from the connection of a 50-ohm feed line to a resonant antenna having a 200-ohm feed-point impedance?
A. 4:1
B. 1:4
C. 2:1
D. 1:2

Answer: _____

Answers

G1A03	D	G3A02	B	G4D09	D	G8A10	B
G1B04	A	G3B07	D	G4E03	A	G8B02	B
G1C08	B	G3C07	D	G5A05	D	G9A04	C
G1D10	C	G4A09	C	G5B16	B	G9B07	A
G2A07	B	G4B02	D	G6A06	C	G9C04	D
G2B02	A	G4C04	A	G7A05	B	G9D09	A
G2C11	B						

APPENDIX 1
RST Reporting System

The RST Reporting System is a means of rating the quality of a signal on a numerical basis. In this system, the R stands for readability, and is rated on a scale of 1 to 5. The S stands for signal strength, and it is rated on a scale of 1 to 9. The T indicates the quality of a CW tone, and its scale is also 1 to 9. The higher the number, the better the signal.

READABILITY

1. Unreadable
2. Barely readable; occasional words distinguishable
3. Readable with considerable difficulty
4. Readable with practically no difficulty
5. Perfectly readable

SIGNAL STRENGTH

1. Faint; signals barely perceptible
2. Very weak signals
3. Weak signals
4. Fair signals
5. Fairly good signals
6. Good signals
7. Moderately strong signals
8. Strong signals
9. Extremely strong signals

TONE

1. Extremely rough, hissing tone
2. Very rough AC note; no trace of musicality
3. Rough, low-pitched AC note; slightly musical
4. Rather tough AC note, moderately musical
5. Musically modulated
6. Modulated note; slight trace of whistle
7. Near DC note; smooth ripple
8. Good DC note; just a trace of ripple
9. Purest DC (If note appears to be crystal controlled, add letter X after the number indicating tone)

EXAMPLE: Your signals are RST 599X. (Your signals are perfectly readable, extremely strong, have purest DC note, and sound as if your transmitter is crystal-controlled.)

APPENDIX 2 - FREQUENCY CHART

FREQUENCY ALLOCATIONS FOR POPULAR AMATEUR BANDS
All in MegaHertz. "X" indicates no privileges.

CLASSES	NOVICE CW	NOVICE PHONE	TECHNICIAN CW	TECHNICIAN PHONE	GENERAL CW	GENERAL PHONE	ADVANCED CW	ADVANCED PHONE	EXTRA CW	EXTRA PHONE
BANDS										
75-80 METERS	3.675 to 3.725	X	3.675 to 3.725	X	3.525-3.750 and 3.85-4.0	3.85 to 4.0	3.525-3.750 and 3.775-4.0	3.775 to 4.0	3.5 to 4.0	3.75 to 4.0
40 METERS	7.1 to 7.15	X	7.1 to 7.15	X	7.025-7.150 and 7.225-7.3	7.225 to 7.3	7.025 to 7.3	7.15 to 7.3	7.0 to 7.3	7.15 to 7.3
20 METERS	X	X	X	X	14.025-14.15 and 14.225-14.35	14.225 to 14.35	14.025-14.15 and 14.175-14.35	14.175 to 14.35	14.0 to 14.35	14.15 to 14.35
15 METERS	21.1 to 21.2	X	21.1 to 21.2	X	21.025-21.20 and 21.30-21.450	21.3 to 21.450	21.025-21.20 and 21.225-21.45	21.225 to 21.450	21.0 to 21.450	21.2 to 21.45
10 METERS	28.1 to 28.5	28.3 to 28.5	28.1 to 28.5	28.3 to 28.5	28.0 to 29.7	28.3 to 29.7	28.0 to 29.7	28.3 to 29.7	28.0 to 29.7	28.3 to 29.7
6 METERS	X	X	50.0 to 54.0	50.1 to 54.0	50.0 to 54.0	50.1 to 54.0	50.0 to 54.0	50.1 to 54.0	50.0 to 54.0	50.1 to 54.0
2 METERS	X	X	144.0 to 148.0	144.1 to 148.0	144.0 to 148.0	144.1 to 148.0	144.0 to 148.0	144.1 to 148.0	144.0 to 148.0	144.1 to 148.0

* Codeless Technician Operators may use only Technician frequencies above 30 MHz.

APPENDIX 3
TABLE OF EMISSIONS

Since 1979, the electronics industry has been using the WARC Emission symbols. In this system, the old 2-character symbols have been replaced with 3-character symbols. The 3-character symbols give more specific information concerning the emissions that they represent. See Appendix 4 on page AP-4.

FIRST CHARACTER

N Emission of an unmodulated carrier
A AM double-sideband
J Single sideband suppressed carrier
F Frequency modulated
G Phase modulation
P Sequence of unmodulated pulses
C Vestigial sidebands
H Single sideband full carrier
R Single sideband reduced carrier
K Sequence of pulses, amplitude modulated
L Sequence of pulse, modulated in width/duration
M Sequence of pulses, modulated in position/phase

SECOND CHARACTER

0 No modulating symbol
1 Digital information-no modulation
2 Digital information with modulation
3 Modulated with analog information
7 2 or more channels with digital info
8 2 or more channels with analog info
9 Combination analog and digital info
X Cases not otherwise covered

THIRD CHARACTER

N No information transmitted
A Telegraphy for aural reception
B Telegraphy for automatic reception
C Facsimile
D Data transmission, telemetry, telecommand
E Telephony
F Television
W Combination of the above
X Cases not otherwise covered

	Traditional	New Symbol
AMPLITUDE MODULATED		
Unmodulated	A0	N0N
Keyed on/off	A1	A1A
Tones keyed on/off	A2	A2A
AM data		A2D
Keyed tones w/SSB	A2J	J2A

	Traditional	New Symbol
SSB data		J2D
AM voice	A3	A3E
Voice w/SSB	A3J	J3E
AM facsimile	A4	A3C
SSB television	A5	C3F
AM television	A5	A3F
FREQUENCY MODULATED		
Unmodulated	F0	N0N
Switch between two frequencies	F1	F1B
Switched tones	F2	F2A
FM data		F2D
FM voice	F3	F3E
FM facsimile	F4	F3C
FM television	F5	F3F
PULSE MODULATED		
Phase	P	P1B

APPENDIX 4
FCC EMISSION DEFINITIONS

The emission designators, given to Appendix 3, were considered, by the FCC, to be too complex. In the reorganization of the Part 97 Rules and Regulations, effective September 1, 1989, the emission designators were grouped into a simple system based on nine terms that are already familiar to amateur operators. This makes it possible for amateur operators to understand immediately their authorized emissions. These nine terms are defined below.

(1) **CW**. International Morse code telegraphy emissions having designators with A, C, H, J, or R as the first symbol; A or B as the third symbol; and emissions J2A and J2B.

(2) **DATA**. Telemetry, telecommand and computer communications emissions having designators with A, C, D, F, G, H, or R as the first symbol; 1 as the second symbol; D as the third symbol; and emission J2D. Only a digital code of a type specifically authorized in this part may be transmitted.

(3) **IMAGE**. Facsimile and television emissions having designators with A, C, D, F, G, H, J or R as the first symbol; 1, 2 or 3 as the second symbol; C or F as the third symbol; and emissions having B as the first symbol; 7, 8, or 9 as the second symbol; W as the third symbol.

(4) **MCW**. Tone modulated International Morse code telegraphy emissions having designators with A, C, D, F, G, H or R as the first symbol; 2 as the second symbol; A or B as the third symbol.

(5) **PHONE**. Speech and other sound emissions having designators with A, C, D, F, G, H, J or R as the first symbol; 1, 2 or 3 as the second symbol; E as the third symbol. Also speech emissions having B as the first symbol; 7, 8 or 9 as the second symbol; E as the third symbol. MCW for the purpose of performing the station identification procedure, or for providing telegraphy practice interspersed with speech. Incidental tones for the purpose of selective calling or alerting or to control the level of a demodulated signal may also be considered phone.

(6) **PULSE**. Emissions having designators with K, L, M, P, Q, V or W as the first symbol; 0, 1, 2, 3, 7, 8, 9 or X as the second symbol; A, B, C, D, E, F, N, W or X as the third symbol.

(7) **RTTY**. Narrow-band direct-printing telegraphy emissions having designators with A, C, D, F, G, H, J or R as the first symbol; 1 as the second symbol; B as the third symbol; and emission J2B. Only a digital code of a type specifically authorized in this Part may be transmitted.

(8) **SS**. Spread-spectrum emissions using bandwidth-expansion modulation emissions having designators with A, C, D, F, G, H, J or R as the first symbol; X as the second symbol; X as the third symbol. Only a SS emission of a type specifically authorized in this Part may be transmitted.

(9) **TEST**. Emissions containing no information having the designators with N as the third symbol. "Test" does not include pulse emissions with no information or modulation unless pulse emissions are also authorized in the frequency band.

APPENDIX 5
ANTENNA ELEMENT EQUATIONS

Yagi Antenna

$$\text{Length (feet)} = \frac{\text{wavelength factor}}{\text{Frequency (MHz.)}}$$

The wavelength factor for: the reflector element = 490
the driven element = 468
the director element = 458

Cubical Quad Antenna

$$\text{Total Length (feet)} = \frac{\text{wavelength factor}}{\text{Frequency (MHz.)}}$$

The wavelength factor for the driven element = 1005
the reflector element = 1030

To get the length of each side divide the total length by 4

Delta Loop Antenna

The wavelength factor for the driven element = 1005
the reflector element = 1030

$$\text{Total Length (feet)} = \frac{\text{wavelength factor}}{\text{Frequency (MHz.)}}$$

To get the length of each side divide the total length by 3

GLOSSARY

Alternating current (AC): Electrical current which reverses its flow every half cycle. The number of cycles per second is the frequency.

Alternator: A device used to generate AC current.

Amateur communication: Non-commercial radio communication by amateur stations. FCC rules say all activities must be personal and without any business interest.

Amateur operator/primary station license: A license issued by the Federal Communications Commission that allows an amateur operator to operate an amateur station. It also indicates the class of privileges.

Amateur operator: A person holding a valid FCC license to operate an amateur station issued by the Federal Communications Commission.

Amateur Radio services: The amateur service, the amateur-satellite service and the radio amateur civil emergency service.

Amateur service: A radiocommunication service for the purpose of self-training, intercommunication and technical investigations carried out by amateurs.

Amateur station: A station licensed by the FCC with the equipment necessary for amateur radio communication.

Antenna: A device that receives or transmits signals over the air.

Antenna switch: A switch used to connect one or more antennas to a transmitter or receiver.

Audio Frequency range: The range of frequencies that can be heard by the human ear, generally 20 hertz to 20 kilohertz.

Automatic control: The use of devices and procedures for station control without the control operator being present at the control point when the station is transmitting.

Automatic Volume Control (AVC): A circuit that maintains a constant audio output volume regardless of variations in input signal strength.

Autopatch: A device that allows amateur radio operators to make telephone calls through a repeater system.

Backscatter: A small amount of signal that is reflected from the earth's surface after travelling through the ionosphere. Backscatter can help provide communications in a station's skip zone.

Beam antenna: An antenna array that receives or transmits RF energy in a particular direction. Usually rotatable.

Block diagram: A simplified outline of an electronic system where circuits or components are shown as boxes.

Broadcasting: Transmissions intended for the general public.

Business Communications: Amateur radio operators are prohibited from making any transmissions that would be considered business or commercial.

Call Book: A published list of all licensed amateur operators, available in North American and Foreign editions.

Call sign assignment: The FCC assigns each amateur station its primary call sign.

Certificate of Successful Completion of Examination (CSCE): A certificate of successful completion allowing examination credit for 365 days.

Coaxial cable, Coax: A concentric two-conductor cable in which one conductor surrounds the other, separated by an insulator.

Control operator: An amateur operator designated by the license of an amateur station to be responsible for the station transmissions.

Coordinated repeater station: An amateur repeater station for which the transmitting and receiving frequencies have been implemented by a licensee above the Novice class.

Coordinated Universal Time (UTC): Sometimes referred to as Greenwich Mean Time, UCT or Zulu time. The time at Greenwich, England. A universal time among all amateur operators.

Critical angle: When radio waves leave an antenna at an angle greater than the critical angle, they pass through the ionosphere instead of returning to earth.

Critical frequency: Radio waves above the critical frequency will pass through the ionosphere and be lost in space.

Cubical quad antenna: A directional antenna designed with its elements in the shape of squares.

Crystal: A piece of quartz that has been ground to a size and shape that will produce natural vibrations of a specific frequency. Crystals are used because they have a high degree of stability.

CW: Continuous wave or CW is another term for the International Morse code.

Delta loop antenna: A variation of the cubical quad antenna with triangular elements.

Dipole antenna: The most common wire antenna. Length is equal to one-half of the wavelength. Can be fed by a coaxial cable.

Dummy load: A device which replaces an antenna as a transmitter's load, designed to eliminate RF radiation during testing or tuning of the transmitter.

E layer: The second lowest ionospheric layer, the E layer exists only during the day.

Effective Radiated Power (ERP): The product of the transmitter (peak envelope) power, expressed in watts, delivered to the antenna, and the relative gain of an antenna over that half wave dipole antenna.

Emergency Communication: Any amateur communication directly relating to the safety of any individuals or the protection of property.

FCC Form 610: It is used to apply for a new amateur license or to renew or modify an existing license. It also serves as the application form for an amateur operator/primary station license.

Federal Communications Commission (FCC): A board of five Commissioners, appointed by the President, with the power to regulate all wire and radio telecommunications in the United States.

Feedline Transmission line: An electrical conductor that connects an antenna to a receiver or transmitter.

Field Day: Annual activity to demonstrate emergency preparedness of amateur operators.

Filter: A device used to block signals at certain frequencies while allowing others to pass.

Frequency: The number of cycles of alternating current in one second.

Frequency coordinator: An amateur radio operator who volunteers to keep records of repeater input, output and control frequencies.

Frequency Modulation (FM): A method of varying the frequency of an RF carrier in response to a modulating signal.

Frequency privileges: The transmitting frequency bands available to the various classes of amateur operators. All privileges are listed Appendix 2.

Fundamental frequency: The operating frequency of an oscillator.

Grid: The control element in a vacuum tube.

Ground: A connection between a device or circuit and the earth.

Ground wave: Radio waves that travel along the surface of the earth.

Hand Key: A simple switch used to send Morse code.

Harmful interference: Interference which degrades, obstructs, or interrupts the operation of a radio communication service.

Harmonic: A radio wave that is a multiple of the fundamental frequency. The second harmonic is twice the fundamental frequency, the third harmonic, three times, etc.

Hertz: One complete alternating cycle per second. Named after Heinrich R. Hertz, a German physicist.

High Frequency (HF): The band of frequencies that lie between 3 and 30 Megahertz.

Highpass filter: A device that allows passage of high frequency signals but attenuates the lower frequencies. When installed on a television set, a high-pass filter allows TV frequencies to pass while blocking lower frequency amateur signals.

Ionosphere: Outer limits of atmosphere from which HF amateur communications signals are returned to earth.

Jamming: The intentional malicious interference with another radio signal.

Key clicks, Chirps: Defective keying of a telegraphy signal sounding like tapping or high varying pitches.

Linear amplifier: A device that reproduces a radio signal but at an increased power.

Long wire antenna: A horizontal wire antenna that is typically one wavelength or longer in length.

Lowpass filter: A filter designed to pass lower frequency signals, while attenuating or blocking higher frequency signals.

Major lobe: The pattern of field strength that points in the direction of maximum radiated power from an antenna.

Malicious Interference: Willful intentional jamming of radio transmissions.

MARS: The Military Affiliate Radio System.

Maximum authorized transmitting power: FCC rules state amateur stations must limit the maximum transmitter power to what is necessary to carry out the desired communications.

Maximum usable frequency (MUF): The highest frequency that will be returned to earth from the ionosphere.

Medium frequency (MF): The band of frequencies that lies between 300 and 3,000 kHz (3 MHz).

Mobile operation: Radio communication conducted from a mobile transmitter.

Mode: Type of transmission such as voice, teletype, code, television, facsimile, AMTOR, etc.

Modulate: To vary the amplitude, frequency or phase of a radio frequency wave in accordance with the information to be conveyed.

Morse code (see CW): The International Morse code is interrupted continuous wave communications conducted using a dot-dash code for letters, numbers, etc.

Multiband antenna: An antenna that operates well on more than one frequency band.

Network: A term used to describe multiple packet stations linked together.

No-Code Technician operator: An amateur radio operator who has successfully completed Element 2 and 3A, but not Element 1A (5 WPM code test).

Novice operator: A Novice is an FCC licensed entry-level amateur operator. Novices may operate a transmitter in the following meter wavelength bands: 80, 40, 15, 10, 1.25 and 0.23.

Ohm's law: The basic electrical law explaining the relationship between voltage, current and resistance. The current I in a circuit is equal to the voltage E divided by the resistance R, or I = E/R.

Oscillator: An amplifier with positive feedback. It produces oscillations or vibrations of an audio or radio frequency signal without any input other than DC operating voltages.

Packet radio: A digital method of communicating computer-to-computer. The transmitted information is broken into short bursts. The bursts (packets) also contain addressing and error detection information.

Peak Envelope Power (PEP): 1. The average power supplied to the antenna transmission line during one RF cycle at the crest of the modulation envelope. 2. The maximum power that can be obtained from a transmitter.

Phone patch: Interconnection of amateurs to the public telephone network.

GLOSSARY

Power supply: A device or circuit that will generate the appropriate voltage and current required by another device or circuit.

Propagation: The travel of electromagnetic waves through a medium.

Q-signals: International three-letter abbreviations beginning with the letter Q used primarily to convey information using the Morse code.

QSL card: A postcard sent to another radio amateur to confirm a contact.

RACES (Radio Amateur Civil Emergency Service): A radio service using amateur stations for civil defense communications during periods of local, regional, or national civil emergencies.

Radiate: To convert electric energy into electromagnetic energy (radio waves).

Radio Frequency (RF): The range of frequencies over 20 kilohertz that can be propagated through space.

Radio frequency interference: Disturbance to electronic equipment caused by radio-frequency signals.

Repeater operation: Automatic amateur stations that retransmit the signals of other amateur stations.

RF burn: A flesh burn caused by exposure to a strong RF field.

RST Report: A telegraphy signal report system of Readability, Strength and Tone. See appendix 1

S-meter: A meter calibrated from 0 to 9 that indicates the relative signal strength of an incoming signal at a radio receiver.

Selectivity: The ability of a receiver to separate two closely spaced signals.

Sensitivity: The ability of a receiver to detect weak signals.

Short Circuit: An unintended low resistance connection across a voltage source.

Single-Sideband (SSB): A method of radio transmission in which the RF carrier and one of the sidebands is suppressed.

Skip zone: The area past the maximum range of ground waves and before the range of waves returned from the ionosphere.

Sky wave: A radio wave that is reflected back to earth from the ionosphere.

Spectrum: A series of radiated energies arranged in order of wavelength.

Spurious Emissions: Unwanted radio frequency signals emitted from a transmitter that sometimes cause interference.

Sunspot Cycle: An 11-year cycle of solar disturbances which greatly affects radio wave propagation

Technician: A no-code amateur operator who has all privileges from 6 meters on up to shorter wave-lengths, but no privileges on 10-. 15-. 40-. or 80- meter wavelength bands.

Technician Plus: An amateur operator who has passed a 5-wpm code test in addition to Technician Class requirements.

Telegraphy: Communications transmission and reception using CW International Morse Code.

Telephony: Communications transmission and reception in the voice mode.

Telecommunications: The electrical conversion, switching, transmission and control of audio signals by wire or radio. Also includes video and data communications.

Temporary operating authority: Authority to operate your amateur station while awaiting arrival of an upgraded license.

Terrestrial station location: A location within the earth's atmosphere, including air, sea and land locations.

Traffic: Messages passed from one amateur to another in a relay system.

Transceiver: A radio transmitter and receiver in one unit.

Transformer: A device that changes AC voltage levels.

Transmitter: A device that produces radio frequency signals.

Transmitter power: The average peak envelope power (output) present at the antenna terminals of the transmitter.

Ultra-High Frequency (UHF): Ultra high frequency radio waves that are in the range of 300 to 3,000 MHz.

Upper Sideband (USB): A transmission mode where the carrier and lower sideband have been removed. Amateurs generally operate USB at 20 meters and higher frequencies; lower sideband (LSB) at 40 meters and lower frequencies.

Very High Frequency (VHF): Very high frequency radio waves that are in range of 30 to 300 MHz.

Volunteer Examiner: An amateur operator of at least a General Class level who volunteers his time to prepare and supervise amateur operator examinations.

Volunteer Examiner Coordinator (VEC): A member of an organization that has an agreement with the FCC to coordinate the efforts of volunteer examiners.

Watt: The unit of power in the metric system.

Yagi antenna: A very popular type of directional antenna.

CODE PRACTICE OSCILLATOR & MONITOR IN KIT FORM OR WIRED

MODEL OCM-2 is a solid-state code practice oscillator and monitor that uses the latest IC circuitry. It contains a 3" built-in speaker, headphone terminals, a volume control and a tone control. It is attractively packaged with a two color panel.

With the addition of a few parts, the unit can easily be converted into a CW monitor. It can therefore be used as an operating aid after the code has been learned.

Cat. #OCM-2K - Code Oscillator & Monitor - Kit $19.95
Cat. #OCM-2W - Code Oscillator & Monitor - Wired $24.95

TELEGRAPH KEYS

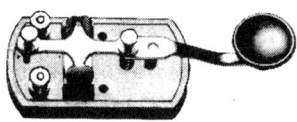

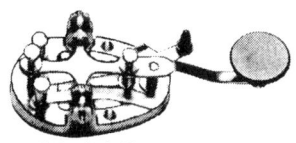

Model K-1   Model K-4

Model K-1 is a low cost, general purpose key with a molded phenolic base and 1/8" contacts. The spring tension and contact spacing are easily adjusted. − $7.95

Model K-4 is a high quality key with a polished brass base, a shorting bar, and 1/8" silver contacts. The contact spacing and tension are easily adjusted. Smooth keying is assured by adjustable ball bearings. − $12.95

STOP TV, FM & VCR INTERFERENCE WITH AMECO HIGH-PASS FILTERS

HP-75T   HP-300T

Eliminate interference from Ham, CB, short-wave, medical equipment, etc. Each filter has 9 shielded sections, a sharp cut-off at 52 MHz. and over 70 dB attenuation below 50 MHz.

Cat. #HP-75T is for coax. feedline and is ideal for the new cable systems. Included are all connectors and a length of coax. − $12.95

Cat. #HP-300T is for twin-lead installations and includes all connectors, terminals and twin-lead with lugs. − $12.95

NEW PREAMPLIFIER FOR TRANSCEIVER USE - MODEL PT-3

- Improves sensitivity while receiving
- No modification required to transceiver
- For single sideband, AM or CW use
- Can handle transceiver output up to 350W.
- Can be used with separate linear amplifiers
- For transceiver or receiver use
- Adjustable delay control on front panel
- Optional second receiver capability
- Optional separate receiving antenna capability

Model PT-3 is a continuously tunable 6-160 meter preamplifier, specifically designed for use with a transceiver. It features a dual-gate FET transistor amplifier which provides a low noise figure, thus improving the sensitivity of the receiver section of the transceiver. Signals are increased as much as 26 dB. A unique built-in RF sensing circuit enables the PT-3 to bypass itself when the transceiver is transmitting. The gain of the preamplifier can be varied by the front panel RF gain control. This control can also be used to reduce the amplification prior to the first mixer, thereby minimizing or eliminating overload effects caused by strong off-channel signals. The PT-3 has an adjustable delay control on the front panel that determines the amount of time the PT-3 will stay in the transmit mode before returning to the receive mode after the operator stops talking.

Provisions are included so that the PT-3 can easily be modified to feed the input of a second receiver. The PT-3 can also be modified by the user so that a separate receiving antenna can be used in addition to the main antenna.

The input and output impedances of the PT-3 are nominally 50 ohms. This matches most types of antennas.

12 volts DC is required to power the PT-3. If 120 volts AC is available, the Ameco Adapter, Model P-12T, can be used. It plugs into the 120 volt AC source and delivers an output of 12 volts DC for the standard PT-3.

Cat. #PT-3 Preamplifier for 160-6 meters. Wired & Tested.
Cat. #P-12T Adapter-120 V.AC to 12V.DC. Wired & Tested.

Cat. #PT-3..........$124.95   Cat. #P-12T............$9.95

AMECO CORPORATION

224 East Second Street - Mineola, New York 11501
TEL: (516) 741-5030 • FAX: (516) 741-5031

STOP TV, FM & VCR INTERFERENCE

with the new AMECO

HIGH PASS FILTERS

HP-300T

HP-75T

Eliminate interference to TV, FM and VCR sets caused by Amateur Radio, CB, shortwave, police, medical equipment etc. with high quality, lab-type filters. Each filter has 9 shielded sections, a sharp cut-off at 52 MHz. and over 70 dB attenuation below 50 MHz.

Model HP-75T is for 75 ohm coax installations. It is also perfect for the new cable systems. Installation is simple because all connectors and a length of coaxial cable are included. $12.95

Model HP-300T is for 300 ohm twin-lead systems. Installation is simple because the filter input has screw terminals to receive the antenna twin-lead and the filter output has twin-lead with spade lugs to connect to the TV set. $12.95

AMECO CORPORATION
224 East Second Street
Mineola, NY 11501
Phone: (516) 741-5030 • Fax (516) 741-5031

NEW!

Version 2.0
COMPLETE MORSE CODE COURSE FOR IBM PC AND COMPATIBLES

- Generates random characters, numbers and punctuation marks at any speed and tone
- Generates infinite random QSO's - similar to FCC/VEC exams - at any speed and tone
- Sends text from any external data file
- Includes 32 page book on code learning and user's manual
- Learn Morse Code the fun and easy way
- Plus many, many more features

This is the most versatile code course ever developed. It has more features and code training capability than all other code courses on the market combined! It is user friendly, menu driven, and contains over 18 options. It will run on any IBM PC/XT/AT or 100% compatible - at any clock speed, in either monochrome or color.

There are many other features, including quiz sessions for the beginner and the ability to alter letter, character and word spacing. The program can also turn the keyboard into a straight or iambic keyer.

This program allows for an infinite amount of code to be received. Random material and combinations are constantly being sent. It is thus impossible to memorize anything.

This course is ideal for the beginner, and perfect for the licensed ham who wishes to upgrade! And it comes from Ameco, the oldest and largest publisher of code training material.

Ameco's Morse Code Course for Computers
Includes both 3-1/2" and 5-1/4" disks
Catalog Number 107-PC $29.95

AMECO CORPORATION

224 East Second Street ▪ Mineola, New York 11501
TEL: (516) 741-5030 ▪ FAX: (516) 741-5031